T0269465

SpringerBriefs in Molecular Science

Green Chemistry for Sustainability

Series editor

Sanjay K. Sharma, Jaipur, India

Angelo Albini · Stefano Protti

Paradigms in Green Chemistry and Technology

 Springer

Angelo Albini
PhotoGreen Lab, Department of Chemistry
University of Pavia
Pavia
Italy

Stefano Protti
PhotoGreen Lab, Department of Chemistry
University of Pavia
Pavia
Italy

ISSN 2191-5407   ISSN 2191-5415   (electronic)
SpringerBriefs in Molecular Science
ISSN 2212-9898
SpringerBriefs in Green Chemistry for Sustainability
ISBN 978-3-319-25893-5   ISBN 978-3-319-25895-9   (eBook)
DOI 10.1007/978-3-319-25895-9

Library of Congress Control Number: 2015953812

Springer Cham Heidelberg New York Dordrecht London

Printed on acid-free paper

Springer International Publishing AG Switzerland is part of Springer Science+Business Media
(www.springer.com)

Preface

During the last decades, green chemistry has emerged as a consistent discipline, with a set of concepts or thought patterns, including theories, research method, and standards. Thus, the social uneasiness about the damage to the environment and human health caused by industry, in particular chemical industry, and about the waste of non-renewable natural resources that has been a main issue from 1950 on has received an answer, or at least the direction to follow has been defined, through a strictly scientific approach. "Green" is no more an eye-catching word added to the title of a paper or a patent whenever wished, as in previous times "new" or "novel", but rather a precise qualification that attests the belonging to a recognized discipline. Green chemistry is certainly interdisciplinary and involves contributions from every part of science (not only chemistry), but the adherence to a firm set of paradigms can be quantitatively assessed through a recognized green metrics. Within the small dimensions proper to this form, this brief wishes to share with anybody interested our view of such paradigms.

We are deeply thankful to colleagues and students in Pavia that contributed in various ways to this text, in particular Prof. Maurizio Fagnoni and Dr. Davide Ravelli, as well as to the Springer team.

Contents

Chapter 1
Introduction

Abstract The growing awareness that a serious damage to the environment had been caused and valuable resources consumed stirred up the social and political conscience and preservation of the environment became one of the key issues in the political arena during the 1960s. What impressed more common people was the destruction of the ecosystem by strongly toxic compounds used e.g. as pesticides in agriculture or for new commodities, e.g. polymers. Large amounts of aggressive products and intermediates were buried in chemical plants, and leakage from such deposits may cause pollution of water bodies. In the following decades, the reversed point of view was introduced, focusing on a new way of preparing useful chemicals while avoiding to produce toxic by-products. This was tagged Green Chemistry and found a meaningful expression in the 12 green chemistry principles.

Keywords Environmental pollution · Sustainability · Green chemistry · 12 principles

1.1 Preserving the Environment

Fireflies. During the early 1960s fireflies began to disappear because of air pollution and, particularly in the country, of water pollution (no more blue rivers and clear water in brooks). "This was a fast and stunning event, and after a few years, the insects became a rather excruciating souvenir from the past. Thus, an old man who remembered their presence couldn't recognize in new young people what he had been as a young man and was deprived even of such a regret" [1]. The Italian poet Pierpaolo Pasolini commented the disappearance of fireflies as a consequence of the large use of insecticides and weed-killers. He took this date as a turning point, where also a reversal of old social values had occurred [1].

This was certainly not the first time that similar issues were mentioned. Indeed, the history of man is also the history of the reckless exploitation of the environment and the 10 thousand years during which human beings have been the dominating

© The Author(s) 2016
A. Albini and S. Protti, *Paradigms in Green Chemistry and Technology*,
SpringerBriefs in Green Chemistry for Sustainability,
DOI 10.1007/978-3-319-25895-9_1

species have produced a consistent series of environment destructions and disasters. Future generations have thus been robbed of a large part of the opportunities previous generations had found available. However, the fast development of chemical industry (and of industry in general) in the last 150 years has made these changes proceed at a fast increasing pace and awaken in the contemporary man the conscience that there was an actual risk that evolution couldn't be controlled and that the damage created was irreversible. Industry is now concerned with environmental issues, but its attitude had not been as open-minded before. Many times a vital choice has been arrived at on the basis of the maximal short-time profit only, despite the unquestionable awareness of the risks that such decision involved.

An emblematic case. Consider the production of tetraethyl lead, the antiknock agent that made the skyrocketing expansion of the car market possible in the 1920s. Several leading scientists warned General Motors (GM) scientists of the serious menace to human health and environment that this compound caused, due to the toxicity (which gained to it the nickname of "loony gas") and the risk of explosion [2]. Actually, even as GM and Standard Oil were about to form a partnership and greatly expand the role of this organometallic compound in the gasoline market, its adoption was still questioned because of the risk implied. However, an internal note of the firm in March 1923 left no doubt upon *why* going forward was that important. "The way I feel about the Ethyl Gas situation is about as follows: It looks as though we could count on a minimum of 20 % of the gas sold in the country if we advertise and go after the business—this at three cent gross to us from each gallon sold. I think we ought to go after it as soon as we can without being too hasty ..." [3]. As it has been commented, with gasoline sales around eight billion gallons per year, 20 % would represent two billion gallons, and three cents gross would bring in $60 million per year. With the cost of production and distribution less than one cent per gallon of treated gasoline, more than two thirds of this would be annual gross profit [2]. As it turned out, these original figures were modest compared to the market success that would come later. Furthermore, a bad accident occurred in a plant when production was beginning and caused five casualties (to be added to those due to toxicity), raising public attention and harsh polemics in the papers, but General Motors and Standard Oil insisted that tetraethyl lead had no alternative as an antiknock agent, and, in the absence of a determined resistance by the Federal Institutions, they had this compound dominate the market by the end of the 1920s. The situation did not change until at the beginning of the 1970s it was recognized that combustion of tetraethyl lead produced particles of the metal and this was dangerous for the public health, so that after a phase down, it was banned.

A study based on GM files reveals a second aspect of the controversy involving the auto industry's long term fuel strategy. The main target for the Research Manager, C. F. Kettering, was protecting GM against oil shortages (then expected to occur by the 1940s or 1950s). "His strategy was to raise engine compression ratios with tetraethyl lead specifically to facilitate a transition to well known alternative fuels (particularly ethyl alcohol from cellulose). However, Kettering lost an internal power struggle with GM and Standard Oil Co. Kettering's strategy was discarded when oil supplies proved to be plentiful and PbEt$_4$ turned out to be profitable in the mid-1920s" [3]. Further antiknock additives were considered, such as iron carbonyl (in afterthought not a good choice) and methanol or ethanol, but despite the many counter indications this remained the cheapest additive. Thus, less toxic additives from renewable sources were discarded and production began. Attention to the troubles connected with tetraethyl lead use was revived only at the time of the following oil crisis in the 70–80s, when another "gasoline famine" started up again interest in the matter. As the leaded gasoline crisis abated in the fall of 1925, Kettering noted that the search for a substitute for petroleum had become problematic: "Many years may be necessary before the actual development of such a substitute," he said [4]. Many such cases may be quoted, e.g. the Eternit company that produced asbestos-based materials in several Italian towns (Casale Monferrato, Broni, Bagnoli), while concealing most of (known) risks related with this compound [5].

Beetles and robins. In a highly successful book, Silent Spring (published in 1962), the marine biologist Rachel Carson [6] showed that the systematic use of insecticides and weed-killers seriously contributed to destroy local ecosystems. "Was it wise or desirable to use substances with such a strong effect on physiology for the control of insects, especially when the control measures involve introducing the chemical directly into the body of water?" [7]. The food chain multiplied the effect of such toxic compounds, even when applied in a very low concentration. "Did we really want a world with no birds singing in the morning and serious consequences even on humans for eliminating beetles?" At the time, control agencies did not take too seriously the situation, but biologists feared a catastrophe. Whom had a citizen to believe? Certainly the book by Mrs. Carston greatly helped to change the social attitude, and the preservation of the environment became a key issue of modern social movement, in the US, in Europe and gradually all over the word. A result was the National Environmental Policy Act (NEPA), passed by the Congress in 1969, with the goal of creating and maintaining "conditions under which man and nature can exist in productive harmony." The U.S. Environmental Protection Agency (EPA) was established the following year, and began its activity by banning the use of DDT and other chemical pesticides.

The existence of dismissed plants where chemical waste had been buried by companies over the previous decades and the leakage from the barrels, with potential pollution of drinking water was another main topic [8]. As it appears from the above, up to this moment the focus was on dismissing production of highly toxic chemicals and cleaning up those that had been spread over the planet in previous decades. In every case, early regulations concerning the environment were adopted not so much on the basis of scientific reports, but rather due to the strength of societal activism that demanded a more environment-respectful approach by industry. Worries about the effect of man-made chemicals continued to increase, with the identification of more subtle effects. As an example, many chemicals have an endocrine-disrupting effect that is revealed only when tested under natural ecological conditions, which makes it doubtful whether consistent and definitive data on toxicology will ever be available [9].

However, the new generation that had been educated in a time where environmental awareness was growing took a more innovative path and began to think whether it was a better choice to prevent pollution rather than have to remediate afterwards. In 1983, an independent committee headed by Norwegian Prime Minister Gro Harlem Brundtland was established within the U.N. General Assembly. The main goal of such World Commission on Environment and Development was to examine global environmental issues up to the year 2000, while reassessing critical problems and formulating realistic proposals for solving them. The results of this work was the Our Common Future agenda (1987), where new, non environmental harmful, policies on which the growth of economies and societies should be based were proposed [10]. The Brundtland Commission expressed the first definition of sustainable development as "a process of change in which the exploitation of resources, the direction of investments, the orientation of technological development and any institutional change were all in harmony", thus enhancing "both current and future potential to meeting human needs and aspirations" [10a]. The first symposium on Green Chemistry was organized in January 1993 in collaboration with the National Science Foundation (NSF) and the Council for Chemical Research (CCR). Most of the key steps in the evolution of green chemistry are related to the US Environmental Protection Agency (US EPA) activity. During the 1980s pollution prevention became the priority, instead of end-of-pipeline control, leading to the approval in 1990 of the Pollution Prevention Act by the American Congress [10b]. Specialized groups, such as the Office of Pollution Prevention and Toxics, established within the EPA in 1988, and international bodies such as the Organization for Economic Co-operation and Development (OECD) favored a co-operative effort for improving existing chemical processes and introducing pollution prevention. These first common efforts led to the United Nations Conference on Environment and Development (Rio de Janeiro, 1992) during which the "Rio declaration of environment and development" and the Agenda 21 action program were formulated [11]. As remarked by Metzger, scientists were urged to contribute to the best of their strength to the Agenda 21 program (see for instance chapter 35 where is highlighted "the role and the use of the sciences in supporting the prudent management of the environment and development for the daily survival and future development of humanity") [12]. Most issues related to chemistry that were introduced in various parts of Agenda 21 will be examined in the following chapters. In general, the most important breakthrough arising from the work by both the Brundtland commission and the Rio meeting is the awareness that the goal should be pollution prevention rather than pollution remediation, and the first attitude should be recognized as the most effective strategy for environmental issues. The US Environmental Protection Agency (EPA) staff coined the motto "benign by design" and the phrase "green chemistry".

1.2 Sustainable and Green Chemistry

In a recent paper Warner and coworkers remarked the excessive and sometimes misleading use of terms such as *"green"* and *"sustainable"* in media and even in scientific discussions [13] and advocated that both concepts of "sustainable chemistry" and "green chemistry" were more clearly defined. In particular, the former is a subsystem of sustainability, and represents a broad concept that deals with all aspects of making and using materials and chemical compounds in the man-built world, including safety and risk policy, remediation technologies, water purification, alternative energy, and, obviously, green chemistry. In the definition by Warner, the last discipline is "application agnostic in that it focuses on the building blocks that go into the ultimate technology" [14]. Green chemistry is thus involved in the optimization of synthesis, in the use of renewable rather than non renewable resources (both chemicals and energy) and in the qualitative and quantitative control of artificial materials employed and produced (as well as of the accompanying waste) [13].

As a consequence, green chemistry involves rethinking chemical reactions and chemical processes in such a way that scientists and engineers are enabled "to protect and benefit the economy, people and the planet by finding creative and innovative ways to reduce waste, conserve energy, and discover replacements for hazardous substances". Thus, the target is not limited to control chemical toxicity and includes "energy conservation, waste reduction, and life cycle considerations such as the use of more sustainable or renewable feedstock and designing for end of life or the final disposal of the product" [15]. From the 1990s, some chemistry practitioners began to publish studies where the environmental performance of chemical processes was explicitly considered [14]. Later in this decade, the advancement of green chemistry and its best known definition as "designing chemistry for the environment" was presented by Anastas, who commented that this "has been driven by new knowledge" [16]. The philosophy of green chemistry was centered on minimizing waste formation (expressed by the paradigm of atom economy) and by devising mild conditions (most often by catalytic processes). As Clark put it, "the challenge for chemists and others is to develop new products, processes and services that achieve the societal, economic and environmental benefits that are now required. This requires a new approach which sets out to reduce the materials and energy intensity of chemical processes and products, minimize or eliminate the dispersion of harmful chemicals in the environment, maximize the use of renewable resources and extend the durability and recyclability of products in a way which increases industrial competitiveness" [17].

These concepts were offered in a compact and cutting way in the *12 principles of Green Chemistry* that are reported below [18, 19]. Actually, rather than principles ("laws" or "rules"), these should be considered *guidelines* (indications on how to carry out an action).

1. *Prevention*. It is better to prevent waste formation than to treat or clean up waste after that it has been created.
2. *Atom Economy*. Synthetic methods should be designed to maximize the incorporation of all materials used in the process into the final products.
   The two first principles are considered in a quantitative rather than qualitative form (see Green Metrics in Chap. 2).
3. *Less Hazardous Chemical Syntheses*. Wherever practicable, synthetic methods should be designed to use and generate substances that possess little or no toxicity to human health and the environment.
4. *Designing Safer Chemicals*. Chemical products should be designed to maintain their desired function while minimizing their toxicity.
5. *Safer Solvents and Auxiliaries*. The use of auxiliary substances (e.g., solvents, separation agents, etc.) should be made unnecessary wherever possible and innocuous when used.
   The 3rd, 4th and 5th principles concern safety. It should be noticed that despite the reservation expressed in the 3rd principle, the moral attitude must be uncompromising. New protocols that are inherently safer must be designed and there is no excuse for maintaining the old ones when these do not guarantee safety. Likewise, collaboration between toxicologists and chemists is important for designing "safer chemicals in a truly holistic and trans-disciplinary manner through innovative curricular advancements" [20].
6. *Design for Energy Efficiency*. Energy requirements of chemical processes should be recognized for their environmental and economic impacts and should be minimized. If possible, synthetic methods should be conducted at ambient temperature and pressure.
   As it has been pointed out, this is rather a "forgotten principle", despite the key role of energy. The use of consistently more expensive energy must be minimized, and new processes occurring under mild conditions have to be introduced, following the path taught by nature [21].
7. *Use of Renewable Feedstock*. A raw material or feedstock should be renewable rather than depleting whenever technically and economically practicable.
   The goal is converting the biomass into useful chemicals in a manner that does not generate more carbon than is being removed from "thin air". The difference between C(in) from the air and C(out) from the energy used is the carbon footprint ΔC. Ideally, when using Principle #7, the overall carbon footprint should be designed as positive, so that C(in) > C(out) [22].
8. *Reduce Derivatives*. Unnecessary derivatization (use of blocking groups, protection/deprotection, temporary modification of physical/chemical processes) should be minimized or avoided if possible, because such steps require additional reagents and can generate waste. Here again, learning from nature and enzymatic processes is a good choice [23].
9. *Catalysis*. Catalytic reagents (as selective as possible) are superior to stoichiometric reagents.

10. *Design for Degradation.* Chemical products should be designed so that at the end of their function they break down into innocuous degradation products and do not persist in the environment.

    Advances in mechanistic understandings linking molecular features to hazards and degradability will enable a more comprehensive application of green chemistry to control the effect on the environment. Predictive decision-making tools must provide confidence about hazard and risk in a way that is aligned with the timing and magnitude of such decisions and, most importantly, while there is still flexibility to alter a molecular design or product formulation [24].

11. *Real-time analysis for Pollution Prevention.* Analytical methodologies need to be further developed to allow for real-time, in-process monitoring and control *prior* to the formation of hazardous substances.

    Real-time feedback is essential in proper functioning chemical processes in order to avoid losing control of a process [25].

12. *Inherently Safer Chemistry for Accident Prevention.* Substances used in a chemical process should be chosen to minimize the potential for chemical accidents, including release, explosion, and fire. The adherence to the green chemistry principles will result in a scenario that is also safer (see Fig. 1.1) [26]. This depicts the Hierarchy of Safety Controls and highlights "the difference between focusing on the control and the hazard part of the safety definition.

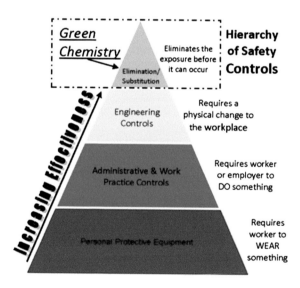

Fig. 1.1 Hierarchy of safety control. Personal protective equipment must be worn every time that it is not possible to fully avoid exposure to a hazard. Administrative and work practice control includes include additional relief workers, exercise breaks and rotation of workers. However, the best precaution is removing the hazard or substituting it with something that is not hazardous, and this is the specific mission of green chemistry. If this is not feasible, engineering control must be applied (enclosing the hazard in such a way that no exposure occurs during normal operations, or at least reduce exposure by a suitable ventilation Reprinted with permission from Ref. [26]

Traditional chemical safety models focus primarily on the control component of that definition"… however… "the most effective means of increasing safety is eliminating the hazard component". Thus, striving towards safer conditions for workers is also striving toward a safer environment for the general public and maintains a safer planet for mankind.

It should further be noticed that green chemistry is by definition close to industrial chemistry, and in particular to chemical engineering, because it is from actual industrial products that comes the main environmental damage, even though small amount of strongly toxic substances occasionally formed may cause major troubles [27]. The purpose of this Brief is to share with researchers and students the paradigms, that is the organized body of notions, on which green chemistry is based. Thus, a small number of recently reported procedures will be presented, with limitation to cases where the environmental performance has been evaluated in a quantitative way. Some issues that have a basic role in this field will be thus considered, namely the use of a recognized *metrics* (Chap. 2), the choice of *renewable* feedstock (Chap. 3), the search for mild conditions for the *activation* of substrates (Chap. 4), the choice of the *solvent* (Chap. 5), the relation between process *intensification* and green chemistry (Chap. 6).

References

1. Pasolini P (1975) Corriere della Sera. Accessed 1 Feb 1975. http://www.corriere.it/speciali/pasolini/potere.html
2. Kovarik B (1994) Charles F. Kettering and the 1921 discovery of tetraethyl lead. Society of automotive engineers, fuels and lubricants division conference, Baltimore, 1994. http://www.environmentalhistory.org/billkovarik/about-bk/research/cabi/ket-tel/#earlyw
3. "Midge" to "My dear Boss" Kettering, 2 Mar 1923, Factory correspondence, unprocessed Midgley files, GMI. Quoted in [2]
4. New York Times, 25 Oct, section 9, p 14. Quoted in [2]
5. Frittoli E (2014) Eternit: la storia del cemento che uccide (1901–2014). http://www.panorama.it/news/cronaca/eternit-storia-cemento-uccide-1901-2014/
6. Carstons R (1962) Silent springs, Houghton Mifflin Harcourt (2002 edition)
7. Page 49 in Ref. [6]
8. van Leeuwen FX (2000) Safe drinking water: the toxicologist's approach. Food Chem Toxicol 38:S51–S58
9. Zala S, Penn DJ (2004) Abnormal behaviour induced by chemical pollution: a review of the evidence and new challenges. Anim Behav 68:649–664
10. (a) Report of the World Commission on environment and development: our common future; UN documents: gathering a body of global agreements. http://www.un-documents.net/wced-ocf.htm. (b) Pollution prevention act of 1990. US Government Printing Office, Washington, 1995, p 617
11. Rio declaration on environment and development, United Nations publication, sales no. E.73. II.A.14 and corrigendum, chapter I. http://www.unep.org/Documents.Multilingual/Default.asp?documentid=78&articleid=1163
12. Metzger JO (2004) Agenda 21 as a guide for green chemistry research and a sustainable future. Green Chem 6:G15–G16

13. Cannon AS, Pont JL, Warner JC (2012) Green chemistry and the pharmaceutical industry. In: Zhang W, Cue BW jr (eds) Green techniques for organic synthesis and medicinal chemistry. John Wiley & Sons, New York
14. Cathcart C (1990) Green chemistry in the Emerald Isle. Chem Ind 5:684–687. Centi G, Perathoner S (2003) Catalysis and sustainable (green) chemistry. Catal Today 77:287–297
15. Linthorst JA (2010) An overview: origins and development of green chemistry. Found Chem. doi:10.1007/s10698-009-9079-4; Clark JH, Macquarrie DJ (1998) Catalysis of liquid phase organic reactions using chemically modified mesoporous inorganic solids. Chem Commun 8: 853–860
16. Anastas PT, Williamson TC (eds) (1996) Green chemistry: designing chemistry for the environment, ACS symposium series, vol 626. American Chemical Society, Washington
17. Clark JH (1999) Green chemistry: challenges and opportunities. Green Chem 1:1–8
18. Anastas PT, Williamson TC, Hjeresen D, Breen JJ (1999) Promoting green chemistry initiatives. Environ Sci Technol 33:116A–119A
19. http://www.acs.org/content/acs/en.html
20. Jimenez-Gonzalez C in Ref. [19]
21. Constable D in Ref. [19]
22. Wool R in Ref. [19]
23. Dunn PJ in Ref. [19]
24. Williams R in Ref. [19]
25. Raynie D in Ref. [19]
26. Bradley S, Finster DC, Goodwin T in Ref. [19]
27. https://www.gov.uk/government/policies/improving-water-quality

Chapter 2
Green Metrics, an Abridged Glossary

Abstract Green chemistry is an aspiration, and the advancement in this field must be recognized and quantitatively assessed. Various proposals of a green metrics have been put forward, based on the consumption of resources, the coproduction of waste, the environmental performance. These are briefly presented, pointing out the specific advantages and limitation of each one. In general, such metrics must blend high level of information supplied with accessibility. Software for several such metrics is freely available.

Keywords Green metrics · Mass metrics · Energy metrics · Environmental metrics · Life cycle

2.1 Environmental Parameters for a Chemical Reaction

What chemists strive to obtain, and what is asked from them, has traditionally been obtaining as much as possible of the desired (saleable) compound. The key parameters have thus been the *reaction yield* (RY) and the *selectivity* (S).

Reaction yield (RY) is the *quantity of a product* (usually expressed as a fraction or a percentage) generated by a chemical reaction from a given reactant. A more correct, but not commonly used praxis should be referring the yield to the balanced chemical equation, thus taking into account the fact that one of the reagents is often used in excess. *Selectivity (S)* is referred to the ratio of one of the products (usually the desired one) arising from the conversion of a certain reactant with respect to the other ones, or to the conversion of the starting material. When a chemical reaction is carried out on industrial scale, the occupation of the available reactors must be taken into account, through parameters such as *productivity* (amount of the desired product per time unit) and *space time yield* (STY), defined as the amount of reaction product formed *per unit volume* of the reactor *per unit time*.

© The Author(s) 2016
A. Albini and S. Protti, *Paradigms in Green Chemistry and Technology*,
SpringerBriefs in Green Chemistry for Sustainability,
DOI 10.1007/978-3-319-25895-9_2

Table 2.1 Current E-Factor value for different industrial sectors from [6]

Industrial sector	Production (Tons year^{-1})	E-factor (kg kg^{-1})	Waste produced (Tons year^{-1})
Petrochemical	10^6–10^8	ca. 0.1	10^6
Bulk chemicals	10^4–10^6	1–5	10^5
Fine chemicals	10^2–10^4	5– 50	10^4
Pharmaceuticals	10–10^3	25–100	10^3

A different issue is having a process that is "green", that is one that causes as little as possible negative effects on the environment. Although good sense will help in judging what will be such effect, specific parameters for the assessment of the environmental performance of chemical reactions have been proposed over the years, with the aim of offering an objective set of metrics for making a process "greener" and making better use both of the materials and of the energy. Proposals have come from various laboratories, sometimes overlapping in some aspects. The metrics are summarily listed below according to their main goals, viz. optimization of the mass used, minimization of environmental damage and of the energy consumed. The most representative parameters are summarized in Table 2.1.

2.1.1 Mass Metrics

As for the mass balance, the parameter *Atom Economy* (AE or atom utilization), has been first defined by Trost in 1991 [1] as "*the ability of a chemical process to incorporate as many as possible of the atoms*" of the starting material into the final products, and thus to the ratio of the molecular weights (MW), see Eq. (2.1):

$$AE = \frac{MW\,(product)}{\sum MW\,(reagents)} \tag{2.1}$$

Convergent syntheses with two or more separate branches can be analyzed by taking into account the amount of the reactants involved in each chemical step, while ignoring the product intermediates [2]. A more elaborated AE expression for multistep synthesis has been proposed by Eissen et al. [3].

A variation, of obvious significance in organic synthesis, is *carbon economy* (CE), proposed by Curzons et al. [4] that is limited to the amount of *carbon* in the reactants that is incorporated in the end product, according to the equation below (Eq. 2.2):

$$CE = \frac{Amount\ of\ Carbon\ in\ product}{Amount\ of\ Carbon\ in\ reagents} \tag{2.2}$$

As originally defined, AE is referred to the chemical equation as such, and thus to a quantitative yield and the use of the reactants in exactly stoichiometric amounts.

Furthermore, neither solvent nor additives (when present) appear in the chemical equation and thus are likewise not considered. This is obviously a significant limitation, and this parameter is better used in conjunction with other metrics. A simple improvement is obtained by considering the yield of the process and introducing a composite parameter, indicated either as the *actual atom economy* (AAE) or as the reaction mass efficiency of the process (RME$_{Kernel}$), defined as the ratio of the actual mass of the products obtained with respect to the reagents used (Eq. 2.3) [5].

$$AAE = RME_{Kernel} = RY \times AE = \left(\frac{mass}{mass}\right) \tag{2.3}$$

This concept can be extended to the *global Reaction Mass Efficiency* (RME$_{global}$ also defined by Sheldon as Material Efficiency, see Chap. 4) that takes into account all of the materials involved in the process, viz. solvents, auxiliaries and chemicals used for the work up procedure. This results in Eq. (2.4), with inclusion of the stoichiometric factor SF for the reagent used in excess, viz.

$$SF = 1 + \frac{\sum mass\ excess\ reagents\ (\text{kg})}{\sum mass\ stoichiometric\ reagents\ (\text{kg})} \tag{2.4}$$

as well as of a *material recovered factor* (MRP) including all of the recovered and reusable materials (starting materials used in excess and recovered at the end of the process, solvents and auxiliaries, see Eq 2.5) [5]. All of these parameters are fractions between 0 and 1.

$$RME_{Global} = AAE \times \frac{MRP}{SF} \tag{2.5}$$

The other way around, one may focus on the concept of waste, which is implicit in the above parameters. Thus, any output from the reaction other than the desired product (that is what is sold) is considered *waste*. The definition thus includes the unreacted starting material, the solvent used (when not recovered), as well as any catalyst or additive, when present. Further to be considered are other products formed beside the desired one, viz. byproducts and coupled products (that is compounds arising from the same pathway that yields the desired product) as well as side-products (that in contrast are produced from the same starting materials used for the synthesis of the target product, but arise from an entirely different mechanism). Furthermore, as mentioned above, often one (or more) of the reactants is used in stoichiometric excess with respect to the other ones. This may well increase the yield of the desired products, but at the same time obviously increases the amount of waste produced. Finally, any purification method used to isolate the product from the crude reaction mixture generates a further amount of waste.

An approach to assess the greenness of a chemical process based on the waste produced was proposed by Sheldon [6, 7] at about the same time as Trost (1992), during the analysis of the industrial production of pharmaceutical intermediates such as phloroglucinol (1,3,5-benzenetriol) [7]. This is the *E-Factor* (E), defined as the ratio between *the mass of waste produced for mass unit of final product* (Eq. 2.6).

$$E = \frac{Mass\ of\ waste\ (kg)}{Mass\ of\ product\ (kg)} \tag{2.6}$$

According to Eq. 2.6, recyclable materials such as solvents, reused reactants or catalysts are not considered as waste and thus ignored, and the ideal value of E is 0. Different parts contribute to the value of the total E-factor (E_{global}). In a detailed analysis, Andraos [8] proposed a more articulated view of this parameter, defined as the sum of different contributions deriving from the core chemical equation (by-products, side-products, and unreacted starting materials, E_{kernel}), from excess reagent (E_{excess}), and from auxiliary materials used in the process, including work-up and purification operations (E_{aux}).

$$E_{global} = E_{kernel} + E_{excess} + E_{aux}. \tag{2.7}$$

As pointed out by Sheldon (see Table 2.2), the value of E-Factor strongly depends on the type of product and on the scale in which it is produced. Thus, in the

Table 2.2 A summary of the main parameters discussed in this chapter

Metrics	Equation	Range of values (ideal value)
Mass metrics		
Reaction yield (*RY*)	$\dfrac{mol\ (product)\ obtained}{mol\ (product)\ expected}$	$0 < RY < 1\ (1)$
Atom economy (*AE*)	$\dfrac{MW\ (product)}{\sum MW\ (reagents)}$	$0 < AE < 1\ (1)$
Reaction mass efficiency kernel (*RME*$_{Kernel}$) or actual atom economy (*AAE*)	$\dfrac{Mass\ of\ product\ (kg)}{Mass\ of\ reagents\ (kg)}$	$0 < RME_{kernel} < 1$ (1)
Reaction mass efficiency global (*RME*$_{Global}$)	$AAE \times \dfrac{MRP}{SF}$	$0 < RME_{global} < 1$ (1)
Environmental factor (*E*)	$\dfrac{Mass\ of\ waste\ (kg)}{Mass\ of\ product\ (kg)}$	$0 < E < \infty\ (0)$
Process mass intensity (*PMI*)	$\dfrac{Mass\ of\ chemicals\ (kg)}{Mass\ of\ product\ (kg)}$	$1 < PMI < \infty\ (1)$
Environmental metrics		
Effective mass yield (*EMY*)	$\dfrac{Mass\ of\ products\ (kg)}{Mass\ of\ non\ benign\ reagents\ (kg)}$	$1 < EMY < \infty\ (\infty)$
E_{IN} (*EATOS*)	$PMI \times Q_{IN}$	$1 < E_{IN} < \infty\ (1)$
E_{OUT} (*EATOS*)	$E\text{-}factor \times Q_{OUT}$	$0 < E_{OUT} < \infty\ (0)$
Energy metrics		
Energy efficiency (E_E)	$\dfrac{Mass\ product\ (kg)}{Energy\ consumption\ (kJ)}$	$0 < E_E < \infty$

case of oil refining, highly evolved (catalytic) systems are used, where waste has been minimized through a long effort. On the other hand, the large volume involved would not make tolerable such processes if this were not the case, both because of the environmental effect and because this would reduce the profit margins. On the contrary, in pharmaceutical industry the tonnage produced is much lower, but the waste produced when preparing (by multistep syntheses) and purifying highly sophisticated materials is much larger, as apparent from the E factor (see Table 2.1).

Yet another approach makes use of the *Process Mass Intensity* (*PMI*, the reciprocal of Reaction Mass Efficiency RME_{global}) as proposed by the Glaxo group [9]. This is defined as *the total mass* of the materials required for the production of the *unit mass of desired product* (Eq. 2.8):

$$PMI = \frac{1}{RME\ global} = \frac{Mass\ of\ chemicals\ (kg)}{Mass\ of\ product\ (kg)} \qquad (2.8)$$

As in the case of E, PMI takes into account the amount of (non reusable) reactants, auxiliaries and solvents employed in the process. In the ideal situation, the PMI value is unitary or close to it (and correspondingly, E = 0). Notice that, as it has been pointed out (see Eq 2.9), E factor and PMI differ by a unity [9]. This is important, since this difference corresponds to the amount of the target product obtained in the process, that is to the actual revenue of the process (see Fig. 2.1). PMI has been considered as a more convenient parameter than E when planning production, because the improvement of the productivity (of the saleable product) and not the waste reduction appears to be a more appealing target. Furthermore, the concept of PMI better matches with the first green chemistry principle of preventing waste production rather than having to find a way to manage it afterwards.

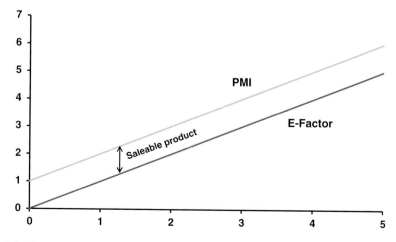

Fig. 2.1 PMI and E-factor differ for one unit which corresponds to the saleable product

$$\text{PMI} = \frac{\text{Mass of chemicals (kg)}}{\text{Mass of product (kg)}} = \frac{\text{Mass of product (kg)} + \text{mass of waste (kg)}}{\text{Mass of product (kg)}}$$

$$= E + 1$$

$$(2.9)$$

Since five parameters (Reaction Yield, the reciprocal of stoichiometric factor SF, AE, RME and the material recovery parameter MRP) well account for the "greenness" of a process, a radial pentagon has been used in order to evidence which are the most sensitive points. Each axis ranges in value between zero and one and in the greenest situation each parameter is equal to 1 (see for review [10]), which results in a regular pentagon. This visual representation has been used by Andraos for the evaluation of different processes, including aldol condensation, Friedel Crafts acylation and cycloaddition (see in Fig. 2.2 an example involving the synthesis of diphenylmethanol via generation of a Grignard reagent and scenarios with different extent of reclaiming excess reagents are evaluated) [11].

It is apparent from the figure that a complete reclaiming is required for a reasonable environmental performance. When applied to a multistep procedure, as typical for Active Pharmaceutical Ingredients (APIs), the use of E-factor in the assessment procedure has the advantage that the contribution for each step is additive, while PMI is not, but is liable to inconsistent application, since the level of solvent recycling, when not measured, is estimated by the evaluator (a 90 % recycling

Fig. 2.2 Synthesis of diphenylmethanol with different extent of reclaiming excess reagent

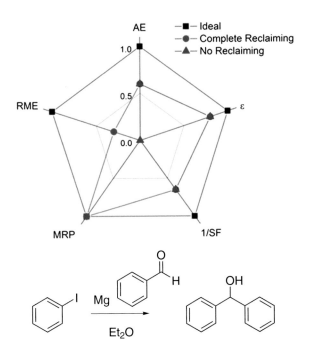

is often assumed). Thus, the use of a *complete* (cEF) and a *simple E-factor* (sEF) have been proposed by Roschangar et al. [12] as defined in Eqs. 2.10 and 2.11.

$$cEF = \frac{\sum m(raw\ materials) + \sum m(reagents) + \sum m(solvents) + \sum m(water) - m(product)}{m(product)}$$

(2.10)

$$sEF = \frac{\sum m(raw\ materials) + \sum m(reagents) - m(product)}{m(product)}$$

(2.11)

These authors suggests that cEF is applied in the post finalization stage, when optimization of the commercial procedure is being carried out, while at an earlier stage sEF is considered. Further determining is the choice of the starting point. Attention is often given to the steps carried out in house, starting from a purchased raw material, but this leaves out part of the environmental relevant processes. Actually, if a raw material is not a commodity, its synthesis must be considered as done especially for that particular API and included into the evaluation. It has been observed that at present 20–50 % of chemical steps are outsourced during the early development and 30–70 % during the late development or after commercial launch of a product. In order to obtain a fair evaluation, it has been proposed to label as raw materials only those that are offered in the Sigma Aldrich catalogue at a price below 100$ per mol (for the largest offered quantity). In the contrary case, the respective synthesis must be included [12].

The conversion of the raw materials into the usually highly complex API involves several steps, and a first appreciation of the greenness of a synthetic plan may be obtained by checking that the number of chemical transformations required for achieving the final complex structure is reduced. Balan [13] has proposed the concept of ideal synthesis as shown in Eq. 2.12.

$$\% \text{ Ideality} = \frac{no.\ of\ construction\ reactions + no.\ of\ strategic\ redox\ reactions}{no.\ of\ reactions}$$

(2.12)

In order to standardize chemical processes across the pharmaceutical industry, the concept of green aspiration level has been introduced. In this way, one is able to define *SMART* (Specific, Measurable, Ambitious and achievable, Result-based, Time-bound) processes as green chemistry goals for the whole field. A standard *aspirational level* (GAL) is defined with reference to the average parameters of processes examined by the ACS Green Chemistry Institute. Roschangar and co-workers calculated the average values as cEF = 307 kg kg^{-1} for Phase 1 and 167 kg kg^{-1} for commercial projects, and sEF = 167 kg kg^{-1} for Phase 1 and 23 kg kg^{-1} for commercial projects. In average, the number of steps in these processes is 7 with 1.3 chemical transformations per step. Thus, the average complexity per drug target is ca. 9 (7 × 1.3). The transformation GAL (tGAL) is therefore expressed by Eq. 2.13 and the process GAL by Eq. 2.14 [12].

$$tGAL = \frac{(s \, or \, c)EF}{Average \times complexity} \tag{2.13}$$

$$GAL = (tGal) \times complexity \tag{2.14}$$

The relative process greenness is thus defined as indicated in Eq 2.15).

$$RPG = \frac{GAL \, (s \, or \, c)EF}{(s \, or \, c) \, EF} \tag{2.15}$$

A RPG > 100 % shows that the green character of the process is below the average industrial value and would benefit from further optimization.

An increased green character of a new process can be evaluated by the reduction of the EF, taking into account the change in complexity. Roschangar and coll demonstrated that decreasing the amount of waste in the overall process does not imply that the RPG doesn't decrease in every single step [12].

2.1.2 Environmental Metrics

E-factor and PMI are the most convenient (and the most easily calculated) parameters for a first assessment of the sustainability of a process. As shown in the following sections, these parameters are largely used as a benchmark in the literature. The main limitation is that these two parameters consider the mass of chemicals involved as a "lump sum", and no account is taken of the quality of such chemicals and the ecological risks related to them. A first attempt to introduce this issue in green metrics was carried out by Hudlicky [14] with the term *Effective Mass Yield* defined as the fraction of the percentage of the mass of desired product relative to the mass of all *non-benign materials* used in its synthesis, according to the equation below:

$$Effective \, Mass \, Yield = \frac{Mass \, of \, products \, (kg)}{Mass \, of \, non \, benign \, materials \, (kg)} \tag{2.16}$$

This is based on the proportion of the mass of the product that arises from non-toxic materials. "Benign" components are defined as 'the by-products, reagents or solvents that have no known environmental risk associated with them, for example, water, low concentration saline solutions, dilute ethanol, autoclaved cell mass, etc.". However, the subjective definition of benign materials is open to criticism.

To date, the most extensive effort to quantify the risk related to a given process is represented by the *EATOS* (Environmental Assessment Tool for Organic Syntheses) facility. The software, developed by Eissen and Metzger in 2004, takes into account as entries a large number of data, which are however easily available, and evaluates a chemical synthesis through four indices, including the above mentioned mass index (PMI) and environmental factor E-factor, as well as two environmental quotients, the "unfriendliness" parameters Q.

Thus, the Environmental Index Input E_{IN} = PMI × Q_{IN}, is *the Potential Environmental Impact* (PEI kg^{-1}) of chemicals used in the process. The factor Q_{IN}, quantifies the environmental and social costs involved in the use of such chemicals (based upon data generally available from the safety data sheet of the chemicals employed, such as risk phrases, the reclaiming of resources involved, transport information and cost).

Analogously, the Environmental Index Output (EI_{OUT} = E × Q_{OUT}) is *the Potential Environmental Impact* (PEI kg^{-1}) on the ecosystem by the chemicals produced. Q_{OUT} is calculated from data available in the Material Safety Data Sheet, by using weighting categories such as human toxicity, chronic toxicity, and eco-toxicology. In addition to the mass and environmental indices, the EATOS software affords also the cost involved in the production of the desired product (expressed in € per kilogram of product). The above contributions well account for the environmental effect [15].

Furthermore, an easy to use semi-quantitative assessment method was reported by Van Aken and takes into account six different characteristics of the reaction, that is yield, cost, safety, technical set-up, temperature and feasibility of workup/purification procedures. In this approach, a range of penalty points is assigned to each of these parameters. As far as the safety is concerned, hazard warning symbols are used to quantify the penalty assigned. Such *ECOSCALE* software uses a scale from 0 to 100, the latter figure representing the ideal reaction, which the Authors defined as "Compound A (substrate) undergoes a reaction with (or in the presence of) inexpensive compound(s) B to give the desired compound C in 100 % yield at room temperature with a minimal risk for the operator and a minimal impact for the environment".

The contribution to the safety, health and environmental performances are detailed in the analogous SHE toolbox [16]. For each chemical and reaction step involved in the process, the SHE aspects are classed into 11 effect categories (fire explosion, reaction decomposition, acute toxicity, chronic toxicity and air mediated effects among others). These are not combined into a single index, but the effect of each substance to a given effect category is individually determined. An idea of the application of this method can be gathered from a paper by Hungerbühler and coworkers where a reaction carried out in a pharmaceutical industry is considered. This is the methylation of 8-α-(tert-butyloxycarbonylamino)-6-methylergonin **1** to compound **2**, one of the six steps of the batch synthesis of building block 8-α-amino-2,6-dimethylergolin, see Fig. 2.3 [17].

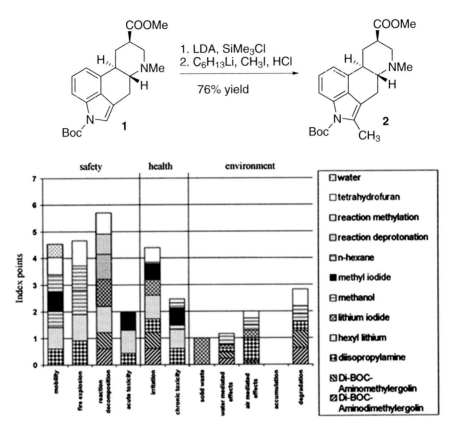

Fig. 2.3 Environmental performance of the materials used in the methylation of 8-α-(tert-butyloxycarbonylamino)-6-methylergolin **1**. Adapted with permission from Ref. [17a]

2.1.3 Energy Metrics

Importantly, none of the above described parameters takes into account the amount of energy supplied for carrying out a chemical process [18]. This is particularly relevant, since in some cases the advantage of using a method that is more eco-sustainable when considering the chemicals used may be lost due to energetic costs. Indeed, in the case of laboratory reactions, the electric energy employed can be measured with good accuracy with easily available, cost-effective energy counters, although only a fraction of the electric energy consumption is actually transferred to the reaction batch. Different energy metrics have been introduced to class a chemical process. One of first parameters defined is the *Energy Efficiency* (E_E) that is the ratio between the amount of the desired product obtained and the electric energy used in a synthesis [18].

$$E_E = \frac{mass\,product\,(\mathrm{kg})}{Energy\,consumption\,(\mathrm{KJ})} \qquad (2.17)$$

Analogously, the *specific productivity* (sP) has been defined as the amount of product (expressed in molar unit, more often used by chemists) obtained for unit of work (in KWh), a definition suitable for any kind of activation, including irradiation by lamps or microwave [19].

$$sP = \frac{mol\,product\,(\mathrm{mol})}{Electric\,work\,(\mathrm{KWh})} \qquad (2.18)$$

The reverse of E_E is defined as the *Energetic process expenditure* (A_P). Energy is consumed both during the reaction (energetic reaction expenditure, A_R) and during work up (energetic work up expenditure, A_E), but often the latter contribution is larger.

$$A_P = A_R + A_W$$
$$= \frac{Energy\,consumption\,(Reaction,\,\mathrm{W}) + Energy\,consumption\,workup\,(\mathrm{W})}{mass\,product\,(\mathrm{kg})}$$

$$(2.19)$$

The most used parameter is, however, the Energy-induced methane equivalents that quantifies the energy consumed as moles of methane required to produce the end product. For determining this value, it is assumed that electricity is exclusively obtained from burning methane, with an efficiency, in the power plant, of 43 %. Then, the amount of methane is calculated in mol (1 MJ = 3.052 mol methane). Alternatively, the amount of CO_2 produced for the process can be also easily calculated.

A general, advanced approach is offered by the *Life Cycle Assessment* (LCA), which follows a philosophy *"from the cradle to the grave"*, where every section of the entire life of the product is assessed [20] including raw material supply, each chemical step, the product or service itself, including its final disposal and waste removal. This approach has been known since the early 1970s when only the energy consumption was investigated, but only in the early 1990s the LCA as we know it today started to emerge. Generally, LCA consists of four steps, namely (1) Goal and definition of the scope. (2) Life Cycle Inventory Analysis, where all the mass and energy flows of the process are recorded according to the defined scope. (3) Life Cycle Impact Assessment, where the results of Life Cycle Inventory process is analyzed in view of its environmental impact, including, among others, climate change, ozone depletion, freshwater and marine eutrophication, human toxicity and water depletion. 4) Life Cycle Interpretation, that involves pointing out the most significant issues related to the process and their evaluation. This approach is operated by a dedicate software regulated by the ISO standards (see the ISO 14000 series). The obtained results can be coupled with other environmental and

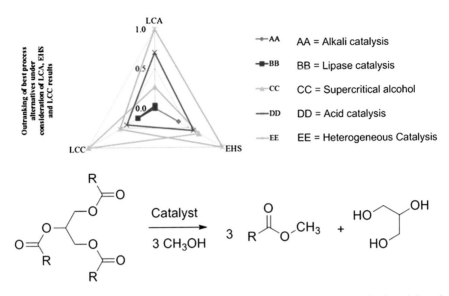

Fig. 2.4 Comparison between different conditions employed for biodiesel production. Adapted with permission from Ref. [21]

evaluation methods in three-dimension graphs, with the aim of affording a complete picture of the process. Kralish et al. analyzed different routes to biodiesel by taking into account the nature of the feedstock, the catalyst (both acid and alkali) and the reaction conditions (heating, microwave activation, use of supercritical fluids), finding that supercritical processing in an intensifying continuous flow reactors is the most favorable proposal (see Fig. 2.4). The results obtained were inserted in a tetrahedral chart including both safety (EHS, see below), environmental (LCA) and economic parameters (Life Cycle Cost, (LCC) that is the analysis of the cost of goods throughout its full life cycle) [21].

In any case, despite the capability of giving a precise idea of both the environmental and (in particular) of the energetic cost of the process, this approach has been substantially limited to a few large scale productions and hardly applied to fine chemistry, because of the fact that the required data are available only for a few chemicals in inventories already present in the LCA database [22].

In order to overcome this limitation, a *Simplified Life Cycle Assessment* (SLCA) has been proposed by the Society of Environmental Chemistry and Toxicology (SETAC), where some chosen approximations (e.g. the use of data of an analogous compounds, rather than exactly of the required one, a move that has been shown often not too largely affect the final result) are applied to the four phases of the traditional LCA [21]. In this way, the LCAs approach is increasingly adopted for the optimization of synthetic routes leading to Active Pharmaceutical Ingredients (APIs) despite the complexity of the problem, in some cases by accepting a reasonable compromise. LCA should offer a realistic measure of the "greenness" of the examined process, be easy to use and able to assess quantitatively the environmental impact of the process, as well as to offer a guidance for the minimization of these impacts. An example is the web-based tool *FLASC*™ (Fast Life cycle Assessment of Synthetic Chemistry) developed at GSK. In this approach, eight impact categories were taken into account, including mass, energy and environmental categories of chemicals such as organic and inorganic reagents and solvent employed in the examined process. FLASC affords, along with RME and PMI values, a score quantifying the LCA impact of the material employed in the process, as well as a quantitative measure of the health effects of the solvent used (EHS) [23]. Further applications of LCA to green chemistry are mentioned in the following chapters [24].

References

1. Li C-J, Trost BM (2008) Green chemistry for chemical synthesis. Proc Nat Acad Sci 105:13197–13202
2. Constable DJC, Curzons AD, Cunningham VL (2002) Metrics to 'green' chemistry—which are the best? Green Chem 4:521–527
3. Eissen M, Mazur R, Quebbemann H-G, Pennemann K-H (2004) Atom economy and yield of synthesis sequences. Helv Chim Acta 87:524–535
4. Curzons AD, Constable DJC, Mortimerand DN, Cunningham VL (2001) So you think your process is green, how do you know?—Using principles of sustainability to determine what is green–a corporate perspective. Green Chem 3:1–6
5. Andraos J (2005) Unification of reaction metrics for green chemistry: applications to reaction analysis. Org Proc Res Dev 9:149–163
6. Sheldon RA (2012) Fundamentals of green chemistry: efficiency in reaction design. Chem Soc Rev 41:1437–1451
7. Sheldon RA (2000) Atom utilisation, E factors and the catalytic solution. C R Acad Sci Paris, IIc, Chimie/Chemistry 3:541–551; Sheldon RA (2007) The E factor: fifteen years on. Green Chem 9:1273–1283
8. Andraos J, Dicks AP (2012) Green chemistry teaching in higher education: a review of effective practices. Chem Educ Res Pract 13:69–79
9. Jiménez-Gonzàlez C, Ponder CS, Broxterman QB, Manley JB (2011) Using the right green yardstick: why process mass intensity is used in the pharmaceutical industry to drive more sustainable processes. Org Process Res Dev 15:912–917
10. Calvo-Flores LFG (2009) Sustainable chemistry metrics. ChemSusChem 2:905–919
11. Andraos J, Sayed M (2007) On the use of "green" metrics in the undergraduate organic chemistry lecture and lab to assess the mass efficiency of organic reactions. J Chem Educ 84:1004–1010
12. Roschangar F, Sheldon RA, Senanayake CH (2015) Overcoming barriers to green chemistry in the pharmaceutical industry—the Green Aspiration Level™ concept. Green Chem 17:752–768
13. Newhouse T, Baran PS, Hoffmann RW (2009) The economies of synthesis. Chem Soc Rev 38:3010–3021
14. Hudlicky T, Frey DA, Koroniak L, Claeboe CD, Brammer LE Jr (1999) Toward a 'reagent-free' synthesis. Green Chem 1:57–59
15. Van Aken K, Strekowski L, Patiny L (2006) EcoScale, a semi-quantitative tool to select an organic preparation based on economical and ecological parameters, Beilstein J Org Chem 2. doi:10.1186/1860-5397-2-3

16. (a) Turney RD, Mansfield DP, Malmen Y, Rogers RL, Verwoered M, Soukas E, Plaisier A (1997) The INSIDE project on inherent SHE in process development and design—the toolkit and its application. ChemESymp Ser 141:202; (b) Koller G, Fischer U, Hungerbühler K (1999) Assessment of environment-, health, and safety aspects of fine chemical processes during early design phases. Comput Chem Eng 23:S63–S66

17. (a) Koller G, Fischer U, Hungerbühler K (2000) Assessing safety, health, and environmental impact early during process development. Ind Eng Chem Res 39:960–972 (http://www.sust-chem.ethz.ch/). (b) Baenziger M, Mak C-P, Muehle H, Nobs F, Prikoszovich W, Reber J-L, Sunay U (1997) Practical Synthesis of 8α-Amino-2,6-dimethylergoline: an Industrial Perspective. Org Proc Res Dev 1:395–406

18. See for review: http://www.oc-praktikum.de/en/articles/pdf/EnergyIndices_en.pdf

19. Ravelli D, Protti S, Fagnoni M, Albini A (2013) Visible light photocatalysis. A green choice? Curr Org Chem 17:2366–2373

20. See for review: Köpffer W (1997) Life cycle assessment from the beginning to the current state. Environ Sci Pollut Res 4:223–228

21. Kralisch D, Staffel C, Ott D, Bensaid S, Saracco G, Bellantoni P, Loe P (2013) Process design accompanying life cycle management and risk analysis as a decision support tool for sustainable biodiesel production. Green Chem 15:463–477

22. Baumann H, Tillman AM (2004) The Hitch Hiker's guide to LCA. An orientation in life cycle assessment methodology and application (swepub.kb.se); Ravelli D, Protti S, Neri P, Fagnoni M, Albini A (2011) Photochemical technologies assessed: the case of rose oxide, Green Chem 13:1876–1884

23. Curzons AD, Jiménez-González C, Duncan AL, Constable DJC, Cunningham VL (2007) Fast life cycle assessment of synthetic chemistry (FLASC[TM]) tool. Int J LCA 12:272–280

24. For further reviews on green metris see: Lapkin A, Constable D (eds) (2008) Green chemistry metrics: measuring and monitoring sustainable processes. Wiley-Blackwell, London, p 344

Chapter 3
Activation of Chemical Substrates in Green Chemistry

Abstract The stability and low polarization of organic molecules forces to use an aggressive chemical or heat to activate (one of) the reagent(s). Addition of an activator worsens the atom economy since spent reagents add to the waste, drastic conditions increase the energetic expenditure. Homogeneous and heterogeneous catalysis, phase transfer catalysis, bio- and photocatalysis, microwave activation, the use of non conventional solvents (supercritical solvents, ionic liquids) or solventless reactions are the means for obtaining a much more environment-friendly process. The application of such methods to various chemical processes is briefly reviewed according to the chemical transformation involved (redox processes, carbon-heteroatom and carbon-carbon bond forming processes), with regards both to commodities and fine chemistry products.

Keywords Catalysis · Phase transfer · Supercritical fluids · Biocatalysis · Microwaves · Photocatalysis · Green reagents · Biosolvents

3.1 Methods of Activation

In 1908 Giacomo Ciamician reflected on the enormous advancement chemistry had achieved in the last decades [1]. It was now safe to state that every natural compound, even the most complex, could be artificially synthesized and showed properties and activity identical to those obtained from natural sources. If something could be reproached to chemistry, however, it was the involvement of brute force to obtain the desired results. Thus, high temperature or pressure and aggressive reagents were required for arriving at the same compounds that plants, for example, synthesized (seemingly) under mild conditions. More than one century later, this issue has been only partially solved and the most important challenge for a synthetic chemist still remains arriving at the target compound in the desired scale, with a (almost) quantitative atom economy, yield and selectivity, under conditions strong enough to allow the reactions to occur at a useful rate, with neither production of

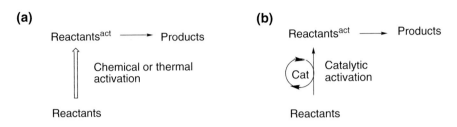

Scheme 3.1 **a** Thermal, (stoichiometric) chemical and **b** catalytic activation of substrates

waste nor excessive energy expenditure. The stability and low polarization of organic molecules forces to use an aggressive chemical or heat to activate (one of) the reagent(s), but if an activator is used, then it is consumed and necessarily the atom economy, the production of waste and the energetic expenditure of the process are affected (Scheme 3.1a). Furthermore, the use of very high temperature to overcome activation energy in chemical processes limits the selectivity. Thus, the use of alternative activation procedures in order to make the process acceptable from both the environmental and the practical point of view is required. The most common approach is the introduction of a catalyst, viz. an activator that is present both among the reagents and the products and is able to activate reactants without being consumed in the process and thus to allow, at least in favorable cases, its use in sub stoichiometric amount in the place of a consumed chemical activator or heat. In catalyzed processes the activation energy of is reduced, which causes an enhancement in reaction rate and/or in the chemo- and stereoselectivity (Scheme 3.1b) [3, 4]. *Catalysis* (defined by Anastas et al. as "a cornerstone in building a sustainable chemical community through green chemistry" [4b]) plays a key role in the improvement in all of the reaction parameters, though advancement in this field is different for bulk and fine chemical production [5]. Thus, ecosustainable catalytic processes have already replaced traditional protocols in bulk chemicals manufacture, whereas in fine chemicals industry stoichiometrical technologies (such as Friedel-Craft acylation) are still used today, even when they involve the generation of large amounts of inorganic waste.

The benefit obtained from the introduction of a catalyst may be very large, arriving at the conversion of a weight of substrate up to ten million times with respect to its own weight [6]. Depending on their state relative to the reaction medium, catalysts are classified into *heterogeneous* and *homogeneous* [6]. In the former case, catalyst and reactants are in the same phase, most often a liquid, with the catalyst dissolved in the reaction medium. On the other hand, a heterogeneous catalyst is usually a solid with the reactant in the gas or liquid phase, and the reaction occurs on the catalytic active surface. The main differences between the two approaches have been highlighted by Lancaster [2] and are summarized in Table 3.1.

In general, the challenge for chemists is to combine the best properties of both choices. In particular, the main limitation related to homogeneous catalysis is the

Table 3.1 Comparing heterogeneous and homogeneous catalysis [2]

	Heterogeneous	Homogeneous
Separation/recover/regeneration of the catalyst	Easy	Difficult/expensive
Reaction rate	Slow	High
Sensitivity to poison	High	Low
Selectivity	Low	High
Service life	Long	Short
Reaction conditions	Harsh	Mild
Reaction mechanism	Often unclear	Well understood

solubility of the catalyst in the reaction medium, that makes its separation from the products and its recover a difficult task for both economical and environmental reasons [7]. Thus, when moving to large scale productions, a "heterogeneization" of the homogeneous catalyst is required. In general, two phase systems offer an elegant approach, with one of the two (often both liquid) phases used as a "mobile-immobilizing" phase for the catalyst, as in the case of *aqueous/organic* system [8] or of the *fluorous/organic* [9] system catalysis.

Actually, *phase-transfer* catalysis (PTC) is one of the most widespread technologies for performing a chemical synthesis under mild conditions while minimizing, at the same time, the production of waste [10]. Generally, it consists in the use of a two liquid phases, an organic phase (where a reaction is promoted by a catalyst, often a commercially available quaternary ammonium salt Q^+X^-) and an aqueous phase, that acts as the reservoir where a reactant (usually an organic or inorganic anion) is regenerated. Alternatively, PTC can also involve both a liquid and a solid phase, in which the catalyst (in most cases a crown ether or a polyethylenglycol) acts as a transport shuttle. Since the phases are immiscible the reaction does not proceed unless the reacting anions are transferred continuously into the organic phase in the form of lipophilic ion pairs with lipophilic cations Q^+, where the other reagents are present. However, PTC is applicable to a limited range of processes, including oxidation and reactions involving anions that are either available as salts (such as cyanide and azide ions) or generated in situ, (such as alkoxides and enolates). Several advantages are related to the use of the PTC approach, including an improvement of productivity (since anions are poorly stabilized in the organic phase their reactivity is amplified in this phase) and selectivity (the reaction proceeds with low activation energy, thus can be run at a lower temperature, lessening the formation of by-products). Furthermore, the use of noxious organic solvents such as *N,N*-dimethyl formamide or hexamethylphosphoramide, otherwise added to the reaction mixture for solubilizing ionic reactants is avoided. Finally, the use of optically resolved catalysts allows the direct synthesis of enantiomerically pure compounds [11]. The use of *supercritical fluids* such as scCO$_2$ [12] or of *ionic liquids* and *deep eutectic* solvents [13] is a further way for modifying the reaction course and simplifying product separation (see Chap. 5).

Among the different catalytic approaches, a key role is performed by *biocatalysis* [14], where biochemical reactions are not only taken as example, but directly exploited in synthesis, by means of natural or in situ produced enzymes. Synthetic approaches that use biological systems are most valuable in terms of the mild conditions adopted and in the amount (and quality) of waste produced. However, even if biocatalytic based processes seem to be per se the most environment-friendly approach in synthesis, the ecological performance of bio-catalysis must be assessed in every single application under well specified conditions. The topic has been extensively reviewed [15]. Another possibility for bringing to reaction a chemical substrate is to supply the energy required for the activation of a (stable) chemical bond in a way alternative to heating. This is the case of emerging technologies involving the use of light energy (photochemistry), ultrasounds and microwaves, whose applications in synthesis will be shortly described in the following sections.

The interest for the use of *microwaves* in synthetic chemistry has greatly increased since their first application in 1986 [16], but many key aspects of their operation are still controversial, in particular related to the application in large scale processes. First of all, reproducibility is an issue, because details of the experimental procedures, as well as of the instrument used, with special regard to the temperature control, are missing in most of synthetic reactions reported in literature. The main limitation however, is that these processes were usually carried out in mmol scale, with a comparatively high power (300–1000 W) radiation. This suggests that troubles may encountered when attempting scale up, in particular in connection with energetic cost [16].

Clark and coworkers [16] compared the energy consumed in carrying out some reactions (including Suzuki couplings, Knoevenagel condensation and a traditional $AlCl_3$ catalyzed Friedel Craft acylation) in one mole scale by using a variety of laboratory facilities, such as traditional oil bath, supercritical CO_2 (sc-CO_2), and microwave reactors. Although Nüchter warned about considering the microwave approach as a universal green method, Clark reported that a significant lessening of the energy requirement was observed in all of the examined processes (up to a 85 fold reduction of the required energy in the case of the Suzuki reaction) [16].

A particular case is represented by *photocatalysis*, where the catalyst is active only after the absorption of a photon (this is particularly desirable when solar light is used). This phrase has been mainly referred to the use of photoactive semiconductors such as TiO_2. This has a band gap energy (3.2 eV) that can be accessed via UV light irradiation and leads to photocatalyzed redox processes [17]. However, this photoprocess is often not selective and is used mainly as an advanced oxidation system for eliminating contaminants from polluted water or as photoactive material for water splitting and CO_2 fixation processes. Quite recently synthetic applications of photocatalysis have experienced a dramatic increase, however [18] (see Scheme 3.45).

In the following, a few examples of synthesis for which the "green" character has been determined by the Authors are reported. Rather than according to the activation mode, these are broadly classified according to the type of process occurring, possibly an easier way to insert these notions in the usual presentation of organic synthesis. Within each group, the structurally simple, high volume products are considered before the fine chemical products.

3.2   Carbon-Heteroatom Bond Formation

In industry, green chemistry has several key issues to confront, from the introduction of oxygenated raw materials in competition with the largely dominating fossil feedstock to the elimination of traces of heavy metals from the catalysts,[1] as well as of toxic or hazardous materials such as organic peroxides or organoselenium compounds and in general any large source of waste. Thus, alternative, less noxious reagents must be considered.

3.2.1   Oxidation/Reduction

Oxidation of alkenes. Due to its use in dilute aqueous solution and to the production of water as the only by product, hydrogen peroxide is considered as a benign oxidant, though it must be activated to interact efficiently with the intermediate concerned. At this regard, PTC can be considered as a promising way to employ H_2O_2. As an example, cyclohexene can be oxidized to adipic acid in 93 % yield with no solvent in the presence of sodium tungstate as oxidation catalyst and methyltricetylammonium hydrogen sulfate as PTC carrier. The solvent-free process, developed by Noyori et al. is an alternative to the conventional synthesis of adipic acid, which involves the emission of 400,000 tons of N_2O per year, corresponding to up to 8 % of the worldwide anthropogenic emission of this gas [19]. Heterogeneous catalytic oxidation processes are widely applied in the bulk chemicals industry, and hydrogen peroxide is one of the elective oxidants for green processes [20]. A good example is the bulk production of ethylene oxide **1**. The conventional chlorohydrin route has an AE of 25 % and an E-factor value of 3 kg kg^{-1}, (the waste is composed by inorganic salts, mainly calcium chloride) respectively, while the direct oxidation of ethylene over a silver catalyst is characterized by a quantitative atom economy and an E-factor value approaching zero (Scheme 3.2).

---

[1]This does not mean that transition metal catalysis has no role in green synthesis. To the contrary, green processes of this type have been reported (although the topic is barely mentioned here for brevity), but trace of metals have to be carefully eliminated, particularly for products used as drugs. See for example: Buchwald SL (2008) Cross coupling. Acc Chem Res. 41: 1439. Liu S, Xiao J (2007) Toward green catalytic synthesis-transition metal-catalyzed reactions in non-conventional media. J Mol Cat A: Chemistry. 270:1–43. Parmeggiani C, Cardona F (2012) Transition metal based catalysts in the aerobic oxidation of alcohols. Green Chem. 14: 547–564.

Scheme 3.2 Oxidation of
ethylene [20]

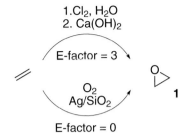

Various oxidation catalysts have been reported [21]. The introduction of Si^{4+} ions with Ti^{4+} in silicalite-1 (TS-1) was early described by Taramasso et al. [22]. The remarkable properties of TS-1 as a catalyst in oxidation processes with hydrogen peroxide aqueous solutions have found a variety of applications in the industry. Enichem has developed a large scale propylene oxide (**2**, PO) manufacture with this catalyst. The reaction is performed in a 30 % aqueous H_2O_2 solution/methanol mixture, with water as the only observed byproduct (97 % yield for the synthesis of propylene oxide) and replaces the chlorohydrin process, thus avoiding the production of inorganic salts (Scheme 3.3) [23].

Among suitable oxidants, atmospheric oxygen is one of the most desirable and participates to reactions either as the ground state molecule or after previous in situ activation to reactive oxygen species, such as singlet oxygen, that causes the desired chemical transformation. Singlet oxygen is a useful reagent in synthesis, as well known since the 1940s, when G.O. Schenck's operated a home-made plant for the sun driven synthesis of ascaridole in the back garden of his house. In this reaction, a diene undergoes a cycloaddition reaction to yield a cyclic endo peroxide by using renewable energy (sunlight), a non toxic solvent (ethanol), as well as a natural pigment (chlorophyll) as photosensitizer [24]. Several other processes via singlet oxygen have been developed.

Synthetically derived flavor rose oxide **4** is a mixture of four 4-methyl-2-(2-methylprop-1-en-1-yl)tetrahydro- *2H*-pyran) diastereoisomers obtained via oxidation of β-citronellol **3** followed by reduction of the resulting peroxides with Na_2SO_3 and acid catalyzed cyclization. The oxidation step is one of the rare industrial applications of photochemistry, and has been patented by Dragoco [25]. The reaction takes place in MeOH in the presence of a dye (rose bengal) as photosensitizer under irradiation (a high pressure Hg lamp). Ravelli et al. carried out an evaluation of different thermal and photochemical routes to rose oxide by means of EATOS and LCA analyses, highlighting the advantages of photosensitized

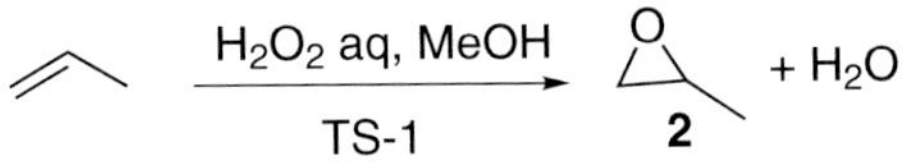

Scheme 3.3 Synthesis of propylene oxide **2**

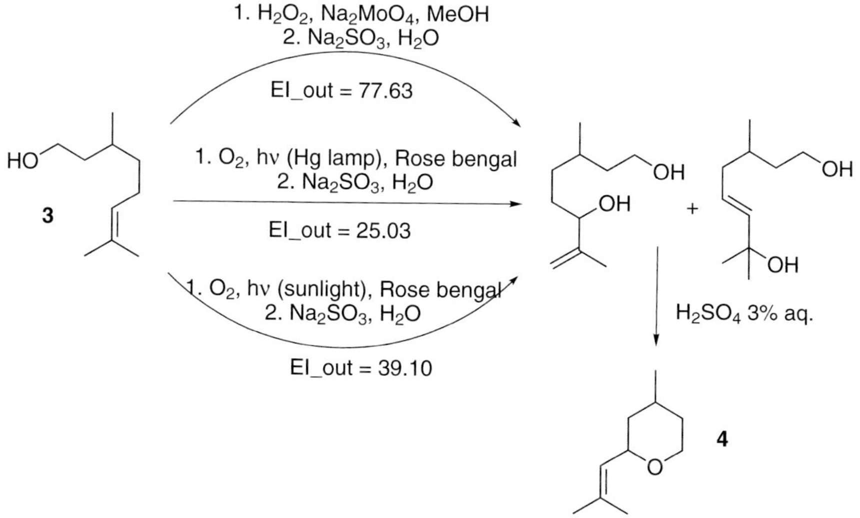

Scheme 3.4 Environmental performance of different processes leading to rose oxide [25, 26]

oxidations compared to the use of a thermal catalyst such as Na$_2$MoO$_4$ (with H$_2$O$_2$ as oxidant). As expected, the large scale optimized Dragoco patent give excellent results in term of waste production, but surprisingly when solar concentrators (PROPHIS) were used as photochemical reactors an energy expenditure comparable with that of Dragoco patent was observed, even if no optimization had been carried out [26] (Scheme 3.4).

Oxidation of alcohols and phenols. A microwave induced solvent free oxidation of alcohols to carbonyls compounds in the presence of recyclable iron(III) nitrate-clay (clayfen) has been described by Varma. This approach eliminates the need for traditional and noxious oxidizing agents such as KMnO$_4$ and CrO$_3$ [27].

The progesterone and corticosteroids precursor bisnoraldehyde **6** has been obtained from the corresponding alcohol by using sodium hypochlorite (NaClO) and a 4-hydroxy-TEMPO as catalyst/co-factor. Noteworthy, the bisnoralcohol **5** is obtained in high yield (up to quantitative) from renewable soya sterol feedstock as a starting material (Scheme 3.5) and the process gave up to 89 % less of non-recoverable organic solvent waste and 79 % less of aqueous waste than previous procedures [28].

A dye sensitized photooxygenation of 1-naphtol and its analogous 5-amido-derivative **7** to the corresponding 1,4-naphthoquinones was carried out under natural sunlight by Oelgemoeller et al. (Scheme 3.6). The desired product **8** was isolated in moderate to excellent yields after 2.5 h insulation, with yields that are mostly higher than those observed under a 500 W halogen lamp, allowing a significant energy savings for solar exposures [29].

Carbonyl to carboxyl. The great potentialities of the previously mentioned catalyst TS-1 in large scale oxidation processes is well illustrated in the synthesis by

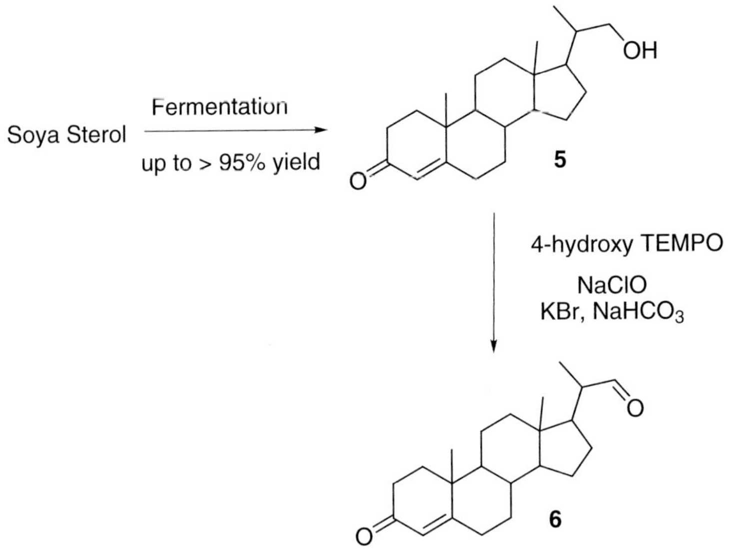

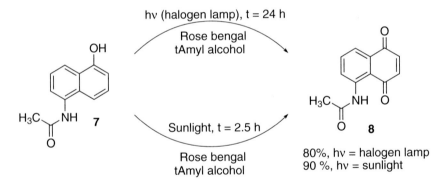

Soya Sterol —Fermentation→ **5**
up to > 95% yield

4-hydroxy TEMPO
NaClO
KBr, NaHCO₃

6

Scheme 3.5 Fermentative synthesis of sterol **5** and mild oxidation to the corresponding aldehyde **6**

hv (halogen lamp), t = 24 h
Rose bengal
tAmyl alcohol

7

Sunlight, t = 2.5 h
Rose bengal
tAmyl alcohol

8
80%, hv = halogen lamp
90 %, hv = sunlight

Scheme 3.6 Photosensitized oxidation (halogen lamp vs sunlight)

oxidative ammonolysis of ε-caprolactame developed by Sumitomo. In this method, the generation of 4.5 tons of ammonium sulfate for ton of desired product arising from the conventional synthesis via hydroxylamine is avoided, and the only byproducts are two molecules of water for molecule of the desired cyclohexenone oxime (**9**, Scheme 3.7) [30].

TS-1 catalyzed ammoximation also finds application in fine chemistry as in the case of p-hydroxyacetophenone (**10**), a precursor of paracetamol **11** that was obtained in 100 % selectivity at 50 % conversion of substrate (Scheme 3.8) [31].

The classic Baeyer–Villiger conversion of a ketone into a lactone moiety has been carried out in the presence of atmospheric oxygen and a genetically-engineered

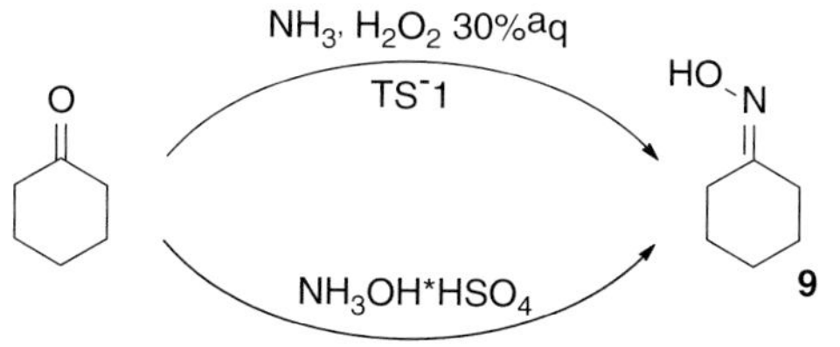

Scheme 3.7 Ammoximation vs hydroxylamine + oxidation

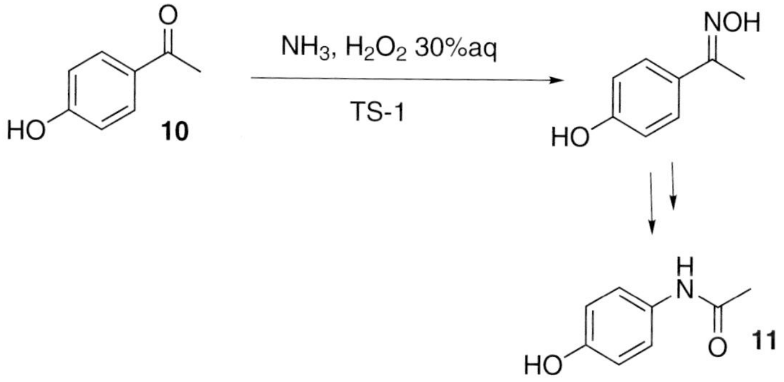

Scheme 3.8 Ammoximation reaction

baker's yeast as biocatalyst, avoiding the shock sensitive and explosive oxidant *m*-chloroperoxybenzoic acid. The reaction is run in an aqueous medium and water is the only by-product formed (an example in Scheme 3.9 for the synthesis of compounds **12**) [32].

Reduction. The biocatalyzed reduction of 2-octanone to (*R*)-2-octanol (**13**) by treatment with alcohol dehydrogenase from Lactobacillus brevis (LdABH) in water in the presence of a ionic liquid as solubilizer was assessed and improved by Leitner and colleagues [33] (Scheme 3.10). In particular, recycling 90 % of the aqueous by product was found to reduce the PMI value from 133 down to 28, with a reduction by 36.4 % of the production cost. By optimizing the enzyme/substrate ratio and the cofactor concentration, PMI was further reduced to 18 kg kg^{-1}.

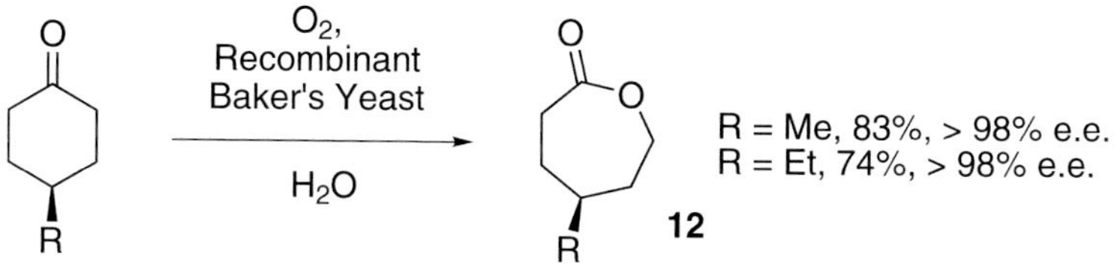

R = Me, 83%, > 98% e.e.
R = Et, 74%, > 98% e.e.

Scheme 3.9 Fermentative Baeyer Villiger oxidation

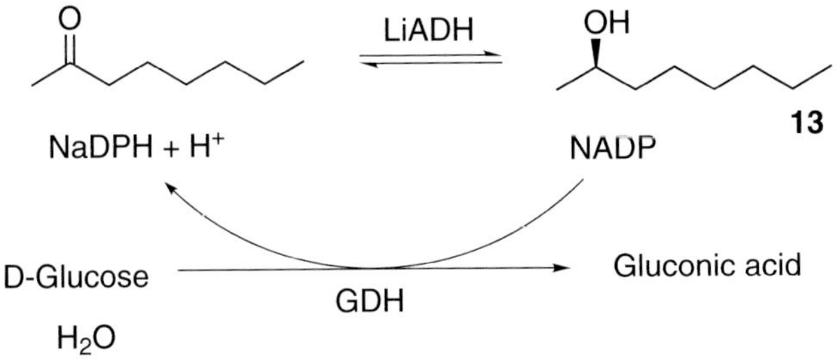

Scheme 3.10 Biocatalytic ketone reduction

In the field of biocatalytic processes, the two-liquid-phase-system (2LPS) concept is one of the most used approaches. The hydrophobic substrate is used either neat or dissolved in a suitable solvent (toluene in most cases) and is added to the second liquid phase, the aqueous, biocatalyst-containing "reactive phase" (Scheme 3.10). The organic phase serves as a substrate reservoir and product sink at the same time. This approach allows to operate with a high substrate loading (this is mainly confined in the organic phase, while the aqueous concentration may remains low), significantly reducing the overall E-factor. An example is the efficient and stereoselective synthesis of ethyl (S)-2-hydroxy-4-phenylbutyrate **15** via reduction of ethyl 2-oxo-4-phenylbutyrate (**14**) by means of recombinant diketoreductase reported by Chen and coworkers (Scheme 3.11) [34]. In the optimized procedure, a substrate concentration of 0.8 M (164.8 g l^{-1}) was used in an aqueous-toluene biphasic system coupled with formate dehydrogenase for the regeneration of the cofactor, resulting in a E-factor value of 8.43 kg kg^{-1}, two orders of magnitude lower than that calculated for the same reaction performed in a single aqueous phase.

Compound LY300164 (**18**, Talampanel) is used in the treatment of epilepsy and neurodegenerative diseases. The first generation synthesis employed safrole as starting substrate and afforded the desired product in poor yield, (16 %), while requiring the use of a huge amount of organic solvents and toxic inorganic oxidant chromium oxide. A chemoenzymatic alternative has been developed by the group of Tao, and involves the biocatalytic reduction of the ketone moiety in 96 % by Zygosaccharomyces rouxii to give the corresponding alcohol (S)-**16**. The reaction was quantitative and the (S) isomer was obtained enantiomerically pure. Ensuing acid-catalyzed coupling with 4-nitrobenzaldehyde and oxidation then led to ketal **17**. The final product LY300164 was obtained after further three steps, in an overall yield of 51 %. Moreover, the new synthetic pathway was largely advantageous in terms of produced toxic waste. Indeed, it avoided the generation of ca 340 l of organic solvents and 3 kg of chromium waste for the production of each kg of **18** (Scheme 3.12) [35].

In the last two decades, thanks to the impressive advancement of biotechnologies, tailor-made enzymes exhibiting high stability and substrate specificity have been applied to several industrial organic syntheses. The key step in the industrial

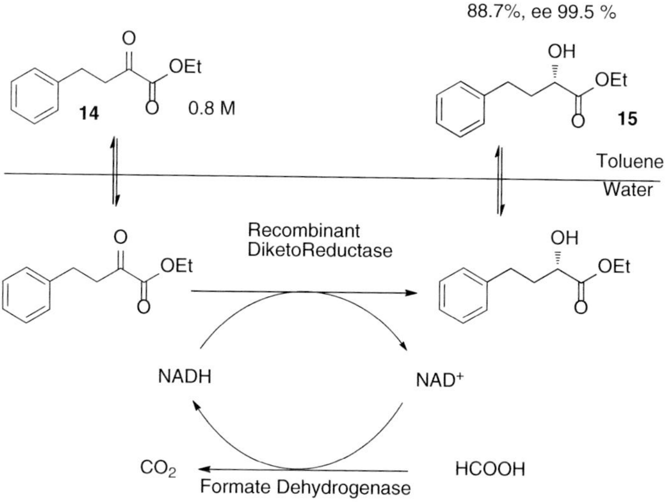

Scheme 3.11 Two-liquid-phase-system

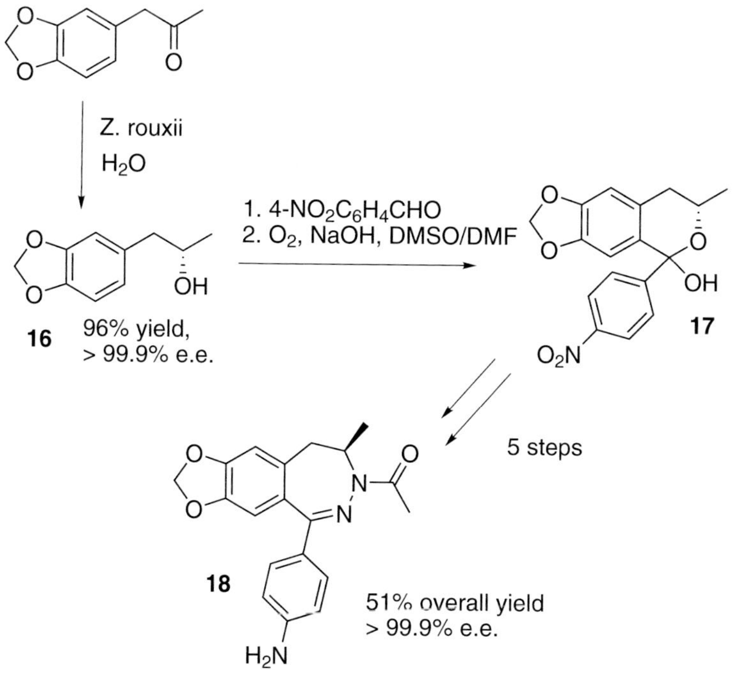

Scheme 3.12 Biocatalytic ketone reduction

synthesis of anti-diabetes drug sitagliptin (**20**, MK-0431 used as the phosphate **21**) is the rhodium catalyzed asymmetric hydrogenation of the corresponding enamine that provides sitagliptin in 97 % e.e. Further purification steps are required for upgrading the e.e. and eliminating rhodium contamination, which, however, affects yield and waste production. An alternative involves the use of a structural homology model of (*R*)-selective transaminase ATA-117. Thus, in the presence of the improved enzyme (6 g l^{-1}), a 200 g l^{-1} solution of prositagliptin ketone **19** in 50 %$_{aq}$ DMSO was converted to sitagliptin **20** in a 92 % yield (>99.95 % e.e.) with an up to 13 % increase in overall yield, a 53 % increase in productivity and a 19 % reduction in total waste [36, 37] (see Scheme 3.13).

3.2.2 N-Acylation and Alkylation

Amidation of acids has been recently recognized by the American Chemical Society Green Chemistry Institute Pharmaceutical Roundtable as a high priority research area. This is one of the reaction classes for which the GSK has compiled a list of reagents according to the environmental issues involved [38a]. A "traffic light" evaluation is assigned to each reagent and reaction on the basis of scores resulting on weighted average evaluations of the effect on environment, health and safety, as well as on atom economy and conditions of reactions and work up, on relevant literature and on internal information (see Fig. 3.1). A "bad note" in any of the parameters considered is reflected in the overall evaluation. Thus, an heterogeneous, readily available, reusable, silica catalyst for the amidation of acids that avoids production of toxic by-products has been proposed by Clark and coworkers [38b].

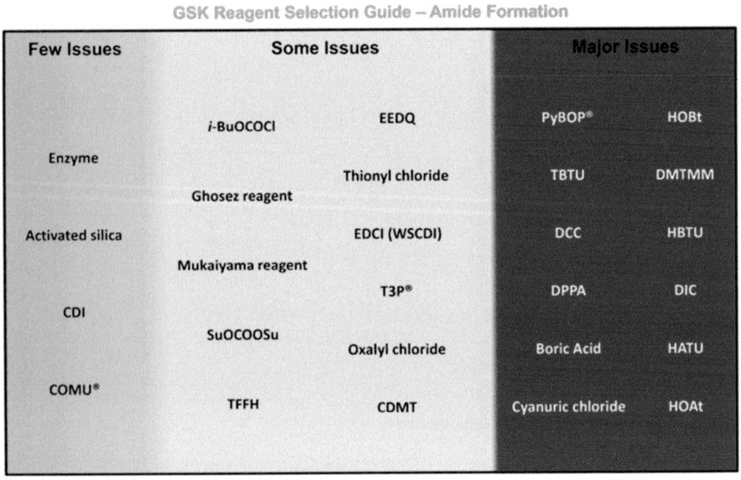

GSK Reagent Selection Guide – Amide Formation

Few Issues	Some Issues		Major Issues	
	i-BuOCOCl	EEDQ	PyBOP®	HOBt
Enzyme		Thionyl chloride	TBTU	DMTMM
	Ghosez reagent			
Activated silica		EDCI (WSCDI)	DCC	HBTU
	Mukaiyama reagent			
		T3P®	DPPA	DIC
CDI	SuOCOOSu			
		Oxalyl chloride	Boric Acid	HATU
COMU®	TFFH	CDMT	Cyanuric chloride	HOAt

Fig. 3.1 GSK amide formation green reagent guide. Reprinted with permission from Ref. [38a]

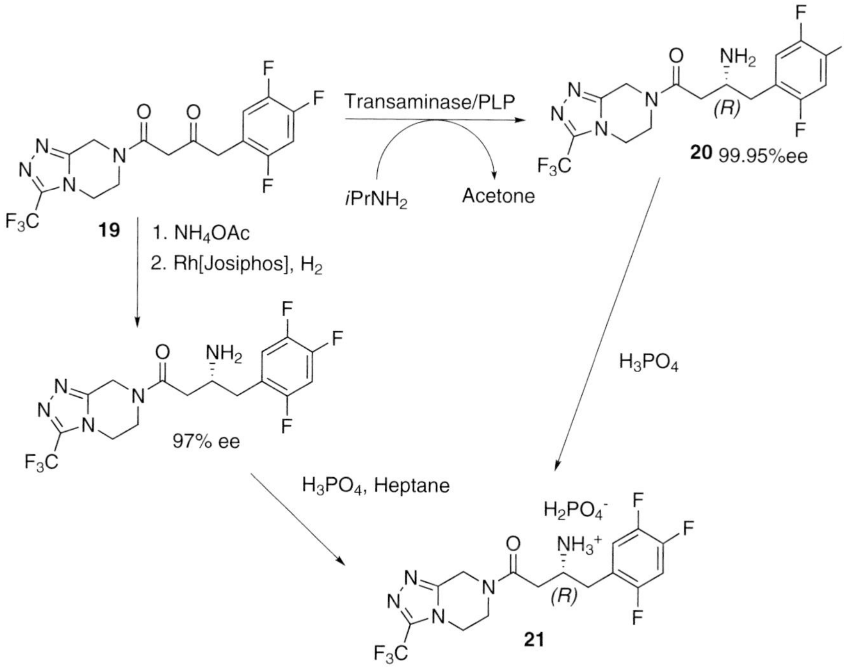

Scheme 3.13 Industrial synthesis of the drug sitagliptin

Thus, K-60 silica activated at 700 °C revealed to be an efficient catalyst for the synthesis of several amides, and neither the use of molecular sieves, nor of Dean Stark apparatus for removing water from the reaction vessel were any more required, thus considerably simplifying the reaction protocol. Noteworthy, when compared to amide synthesis by derivatization by SOCl$_2$ or activation by DCCD, the above solid acid catalysis exhibited environmental metrics improved by an order of magnitude, as shown by the E-factor value that was around 13.5–19.5 kg waste per kg of product in the first cases and dropped down to 1.1 kg waste per kg product in the case of K$_{60}$ catalyst (see an example in Scheme 3.14 for the synthesis of 4,N-diphenylbutirramide **22**).

Solid acid catalysis is increasingly applied [38]. The amidation of esters by using amino alcohols in the presence of 30 % mol potassium phosphate at the relatively low temperature of 60 °C proved effective for the synthesis of several amido-alcohol derivatives in good to excellent reaction yield and with a RME value reaching 85 % [39].

The metal oxide nanocatalysis is considered as a frontier between homogeneous and heterogeneous catalysis because of the large surface area and the possibility of tuning the properties of such materials via control of shape, size, inter-particle spacing and dielectric environment. This approach has been labeled Nanoparticles-catalyzed Organic Synthesis Enhancement (NOSE) and has found several synthetic

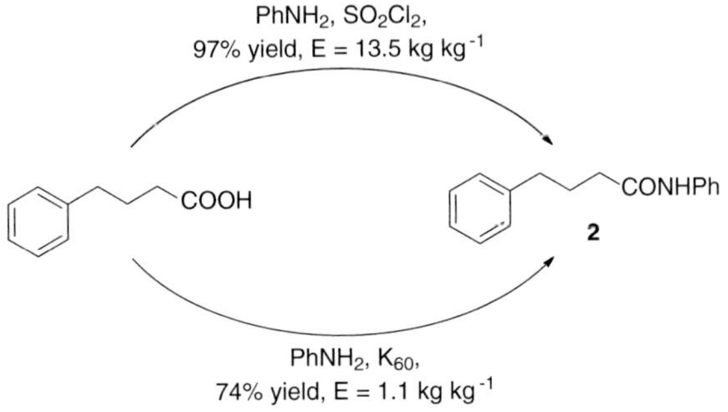

Scheme 3.14 Synthesis of
amide 22

applications. Nano-MgO has been used for the development of a synthetic protocol for the synthesis of amide derivatives under solvent free reaction condition (SFRC) [14, 40, 41]. This has allowed the direct coupling of carboxylic acids with amines with a significant improvement of both versatility and environmental impact with respect to conventional procedures (Scheme 3.15). Again, the E-factor value was lowered by a factor of 15 when passing from chemical functionalization by SO_2Cl_2 ($E = 17$ kg kg^{-1}) to catalysis by MgO nanoparticles (1.07).

The fermentation bulk product penicillin G is a key substrate for the synthesis of different penicillin based antibiotics, including 6-APA (**25**). The traditional route to this target is represented by the Delft-Cleavage (commercialized by the Nederlandsche Gisten Spiritusfabriek) and involves the protection of the carboxyl group in **23** as a silyl ester, followed by the conversion of the secondary amide into the imine chloride **24** by treatment with phosphorus pentachloride and *N,N*-dimethylaniline in dichloromethane, and by hydrolysis, as shown in Scheme 3.16 [42]. The main limitation of the Delft-Cleavage approach is the significant amount of environmentally unattractive reagents including halogenated solvents and reagents, (CH_2Cl_2, Me_3SiCl, and PCl_5 and *N,N*-dimethylaniline) that lead to an E-factor value of 19.2 kg kg^{-1}. This method has been recently supplanted by biocatalytic hydrolysis

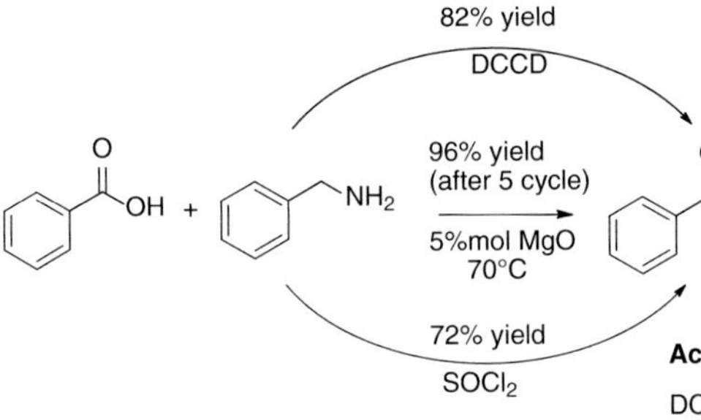

Activant	E-factor kg kg^{-1}
DCCD	22.3
MgO	1.07
SO_2Cl_2	17

Scheme 3.15 Synthesis of amides

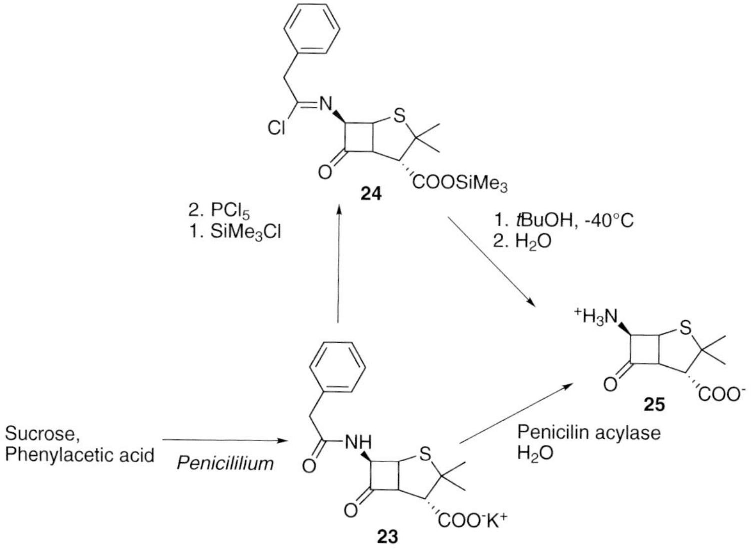

Scheme 3.16 Synthesis of 6-APA

catalyzed by penicillin acylase from *E. coli*, where only 0.09 kg of NH_3 are required for the production of one kg of 6-APA in the fermentation process and water rather than dichloromethane is the solvent chosen for the reaction [43].

Penicillin acylase from *E. coli* is the ultimate catalyst for biocatalytic processes on industrial scale and the synthesis of ampicillin (**26**) from 6-APA (**25**) is one of the best known example. A recently developed suspension-to-suspension reaction system allowed for the overcoming of the main limitation of this process, that is the intrinsically low Synthesis/Hydrolisis (S/H) ratio. This rapidly decreases during the course of the reaction, since the high solubility of ampicillin accelerates its (secondary) hydrolysis. Thus, reactants were added to avoid exhaustion of 6-APA and 0.5 equivalent of *D*-phenylglycine and 1.5 eq. of ammonia were formed in the improved coupling. Interestingly there was no single optimum pH for the coupling reaction, but this depended on the concentration of the reactants. Recycling of the waste by the addition of the chloridrate of *D*-phenylglycine methyl ester as both titrant of generated ammonia and secondary acyl donor resulted in the reduction by a factor of 3 as compared with the standard coupling procedure [44] (Scheme 3.17).

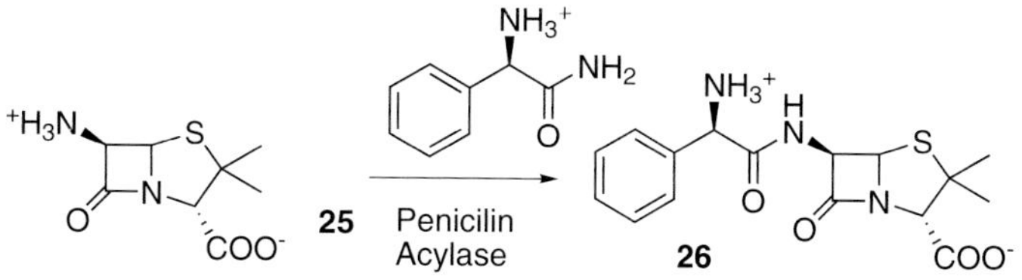

Scheme 3.17 Synthesis of ampicillin

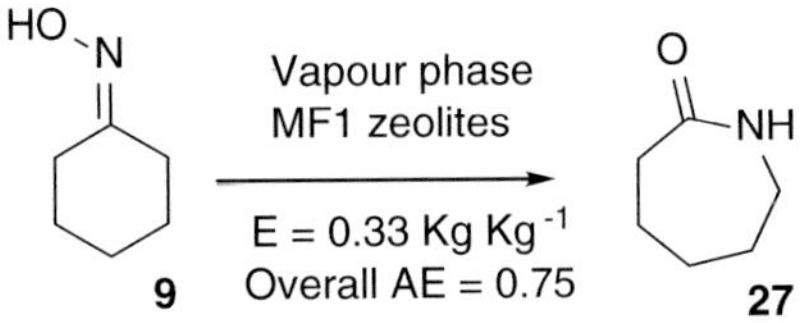

Scheme 3.18 Synthesis of ε-caprolactam **27**

The precursor of nylon 6, ε-caprolactame **27**, can be obtained through a two-step synthesis under heterogeneous catalyzed conditions. Thus, formation of cyclo-hexanone oxime **9** via TS-1 catalyzed ammoximation of cyclohexanone (see Scheme 3.7), is followed by vapor phase Beckman rearrangement on high silica content zeolite MF1 [45]. The process supplanted the traditional oleum induced acid rearrangement, and led to the production of the cyclic amide with an overall AE of 0.75 (including oxime synthesis) and an E-factor of 0.33 kg kg^{-1} (to be compared with E = 2.5 kg kg^{-1} in the case of the conventional path) (Scheme 3.18).

The synthesis of cyclic amines **28** in a scale ranging from 20 mmol to 1 mol was performed under microwave irradiation in open vessels at atmospheric pressure [46]. The target products were obtained under transition metal free conditions, by using water as the solvent. Interestingly, cyclization was found to be complete in 30 min (compared to the 5 h required for the traditional thermal reaction) and purification consisted in a simple water washing of the crude product. Furthermore, moving from a single-mode to a multimode microwave cavities resulted in a scale up of the reaction up to 50 times (from 20 mmol to 1 mol), while no increase of either the time and power of the irradiation was required.

Photocatalysis often allows for the direct functionalization of C–H bonds in reactants, rather than having first to introduce more reactive bonds. Irradiation of alkanes, ethers (acetals) and aldehydes (either aliphatic or aromatic) under Xe lamp or solar light simulated irradiation in the presence of diisopropyl azodicarboxylate (**29** DIAD) and using tetrabutylammonium decatungstate (2 mol%) as the photo-catalyst resulted in a synthesis of C(sp^3)–N bonds in substituted hydrazines (**30**, Scheme 3.19). Furthermore, a three-component acylation of DIAD was developed and led to the building of a C(sp^2)–N bond under a 80 atm CO atmosphere (**31**, Scheme 3.19) [47] (Scheme 3.20).

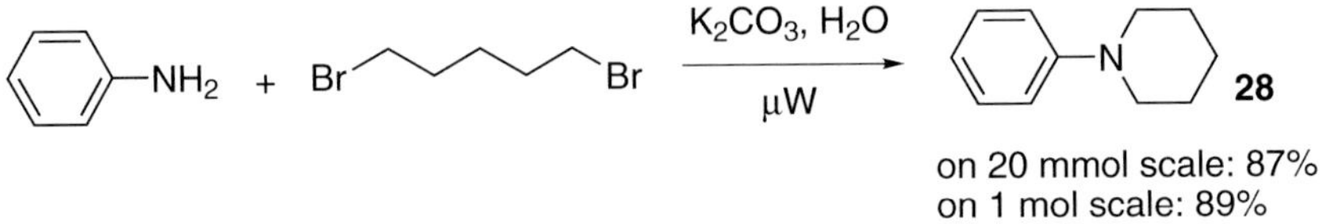

on 20 mmol scale: 87%
on 1 mol scale: 89%

Scheme 3.19 Synthesis of *N*-phenylpiperidine

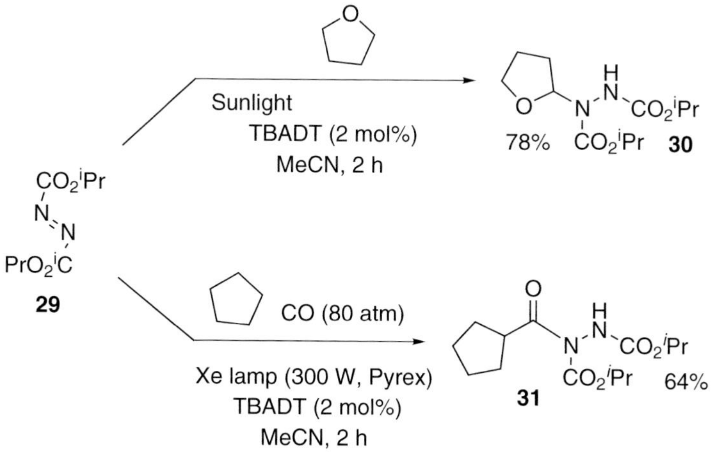

Scheme 3.20 Photocatalitic alkylation of azodicarboxylate

3.2.3 O-Acylation and Alkylation

An assessment of the greenness of different acyl donors in enzymatic acylation has been performed by Paravidino and Hanefeld [48]. This revealed the potentialities of simple carboxylic acids as acyl donors in terms of green chemistry perspective. As an example, the atom economic (AE = 0.97) enzymatic synthesis of polyester **32** from glycerol and adipic acid has been successfully performed on industrial scale under relatively mild conditions (60 °C and 20 mbar) in the presence of Cal–B (Novozym 435) 3 % w/w. Under these conditions a conversion of the starting substrates above 90 % was always achieved with a space time yield of 370 g d^{-1} l^{-1}, an almost quantitative atom economy (97 %) and an extremely low E-factor (0.35), [49, 50] (Scheme 3.21), the largest contribution to the reaction waste being water.

Scheme 3.21 Synthesis of a polyester

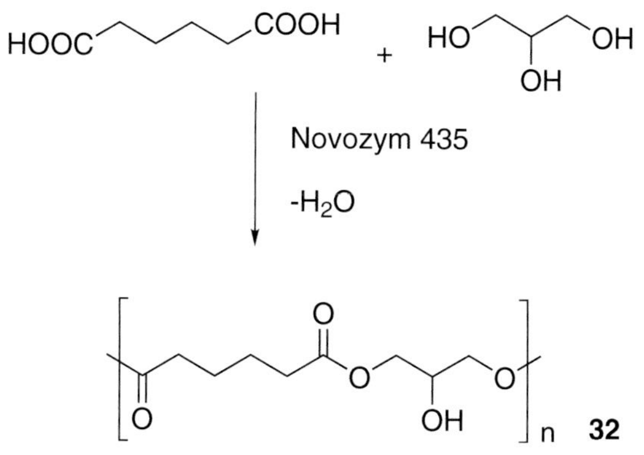

Scheme 3.22 Ethers
preparation: dimethyl
carbonate vs methyl iodide

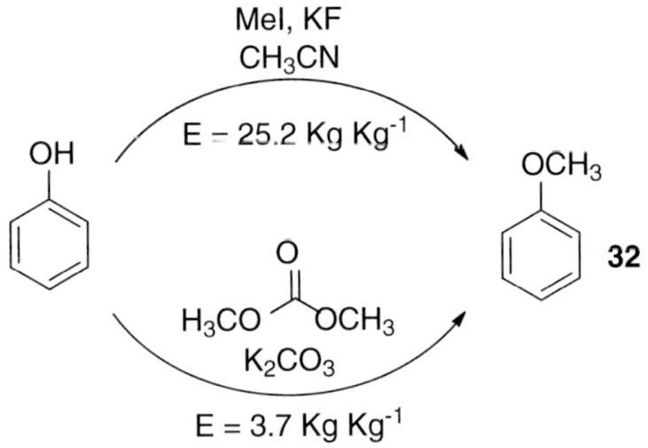

Dimethyl carbonate (DMC) is a benign reagent obtained through phosgene free procedures, e.g. via ZrO_2 catalyzed carboxylation (E = 1.6 kg kg^{-1}) or Cu(I) salts catalyzed oxidative carbonylation of methanol, which can be used avoiding the generation of organic by-products or salts and proceeds with an excellent selectivity when catalysts such as zeolites and K_2CO_3 are employed [51].

DMC has been used as an acylating agent alternative to phosgene in the production of different bulk chemicals including urea, diphenyl carbonate and phenyl isocyanate, the green metrics of which have been assessed by Andraos [52]. However, it likewise functions as methylating agent, with atom economy and mass index values satisfactory when compared to that of traditional reagents such as dimethyl sulfate, methyl iodide and methanol (MeOH) (see an example in Scheme 3.22 for the synthesis of anisole **32**).

Sulfonylation. The synthesis of diarylsulfones (**33**, 11 mmol scale) was carried out in the presence of different Lewis acid catalysts under microwave activation. The most effective catalyst in terms of yield was iron(III) chloride, with a yield of 89 %. Moving from oil bath heating to microwave resulted in one order magnitude reduction of the CO_2 produced during the process (8 vs. 80 kg CO_2/mol) [53] (Scheme 3.23).

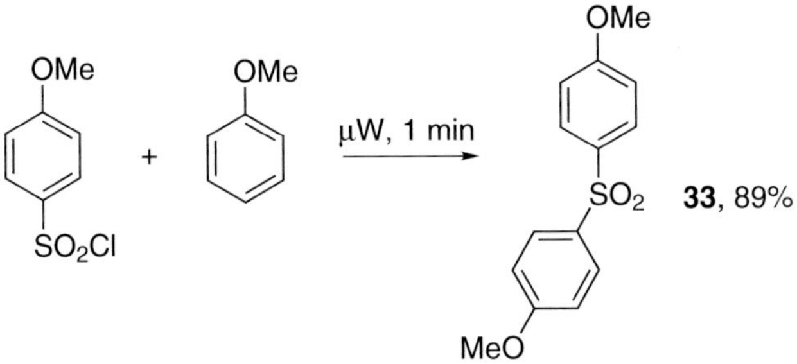

Scheme 3.23 Synthesis of a diarylsulfone

3.3 Carbon–Carbon Bond Formation

Solid acid catalysts are playing a growing role in the field of organic synthesis and the strong points are the reduction of waste and the increased yield. In 2002, a review by Armor showed that 103 industrial protocols involving the use of solid acid catalysts had been reported [54], including, among others, ion-exchanged resin, zeolite and mesoporous (alumino)silicates (the most used materials), clays, oxides and immobilized Lewis and Brønsted acids.

Alkylation. Zeolites are natural or synthetic hydrates aluminosilicates consisting of MO_4 tetrahedra (M = Si, Al) arranged to give well defined micropores, and often interconnected. The use of synthetically derived zeolites allowed for tuning pore size and chemical properties (e.g. hydrophilicity and the type and strength of acid sites) [55]. These are applied to the whole range of acid catalyzed reaction, with the advantage of a closely controlled selectivity. As an example, zeolite Socony Mobile-5 H-ZSM5 (patented by Mobil Oil Company in 1975), is used for the industrial production of a plethora of bulk chemicals including ethyl benzene via alkylation of benzene (Scheme 3.24a), para-xylene (via isomerization of ortho-xylene, Scheme 3.24b) and of cycloxexanol deriving from the hydration of cyclohexene (Scheme 3.24c) [56].

Acylation. A typical case where the key contribution of zeolites in synthesis is highlighted is the *Friedel-Crafts acylation* [57], for which the use of one equivalent or more of a Lewis acid such as $AlCl_3$ or BF_3 is required as the catalyst, producing a large amount of waste. In order to overcome this limitation, β-zeolite has been successful employed in the liquid phase acetylation of anisole by Rhone Poulenc. Thus, whereas in the conventional acylation with acetyl chloride and $AlCl_3$ 4.5 kg of aqueous waste (containing Al salts, HCl and acetic acids) were generated per kg

Scheme 24 Bulk synthesis of ethylbenzene, p-xylene and cyclohexanol

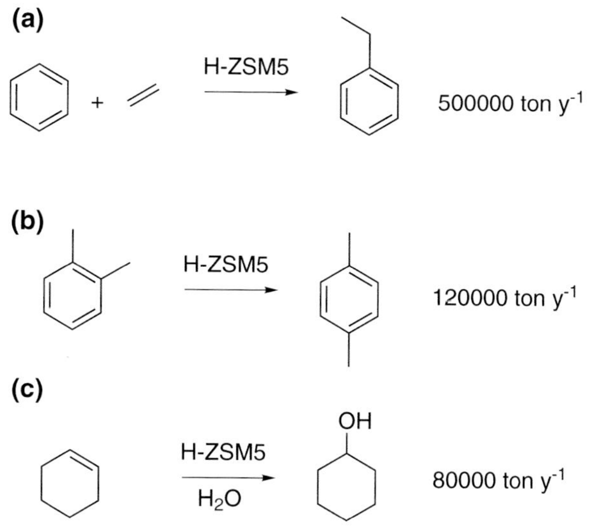

(a)

H-ZSM5

500000 ton y^{-1}

(b)

H-ZSM5

120000 ton y^{-1}

(c)

H-ZSM5

H_2O

80000 ton y^{-1}

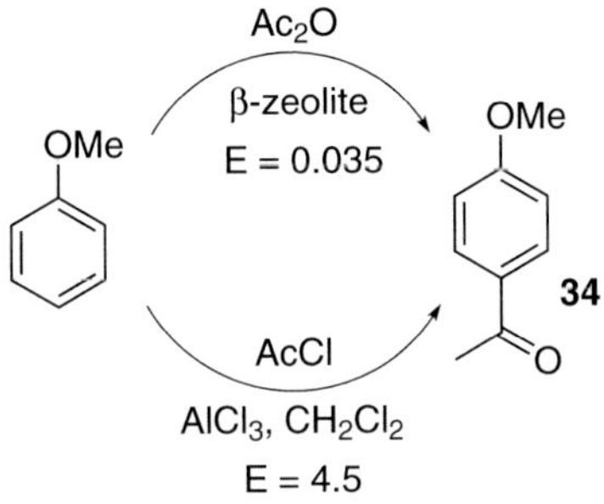

Scheme 3.25 Acylation of anisole

of p-methoxyacetophenone **34**, an E-factor of 0.035 kg kg^{-1} was calculated for the zeolite catalyzed route (the main component of waste was water, Scheme 3.23). Importantly, the catalyst could be recycled and was recovered by either filtration or simple distillation of the products [58] (Scheme 3.25).

The acylation of a phenol followed by Fries rearrangement has been largely used for the synthesis of aromatic ketones. Once again, it proved possible to replace conventional metal chloride catalysis by heterogeneous catalysis employing a solid acid. The synthesis of 2,4-dihydroxyacetophenone **35**, precursor of the UV-absorber 4-O-octyl-2-hydroxybenzophenone (**36**) has been carried out starting from resorcinol and benzoic acid, in the presence of beta H-zeolite. The heterogeneous acid was a less efficient catalyst than FeCl$_3$ usually employed in the conventional synthesis of 4-O-octyl-2-hydroxybenzophenone **36** via benzotrichloride, but this method was advantageous as a chlorine free route. Furthermore the catalyst was regenerated by simple air burn-off (see Scheme 3.26).

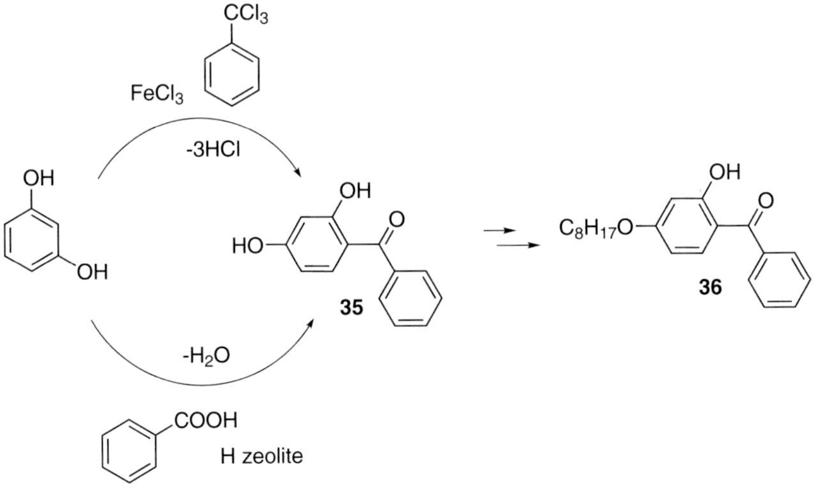

Scheme 3.26 Acylation of resorcinol

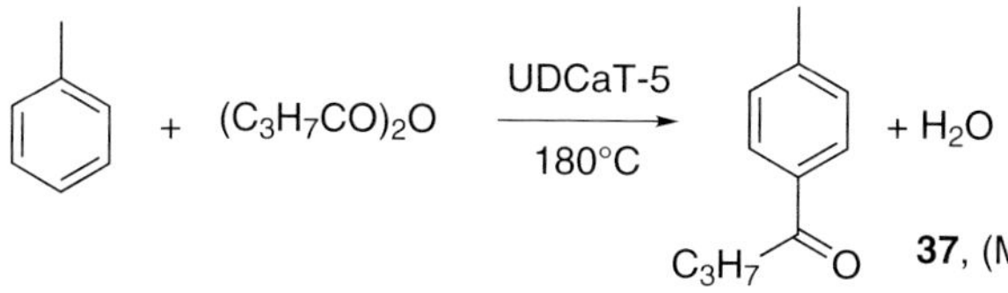

37, (Major isomer)

Scheme 3.27 Acylation of toluene

Zirconia based solid superacid namely UDCaT-4, UDCaT-5, UDCaT-6 found application in various acid catalyzed reaction [59]. The efficient acylation of toluene with propionic anhydride to give 4-methylpropiophenone (**37**) as the major isomer in the presence of UDCaT-5 was optimized by Yadav et al. Thus, a 5:1 toluene propionic anhydride mixture in the presence of 0.06 g mL catalyst underwent a 62 % conversion of priopionic anyhydride, with a 67 % yield of the desired isomer and no di-acylated product (Scheme 3.25). The propionic acid generated likewise reacted with toluene, thus water was the only observed waste [60, 61] (Scheme 3.27).

Metal organic frameworks (MOFs, also known as porous coordination polymers or PCPs) are nanoporous materials constructed from metal containing nodes and organic linkers [62]. Their peculiar structural properties place them at the frontier between zeolites and surface metal–organic catalysts. They have received increasing attention owing to their potential application in chemistry of materials and in synthesis. In particular, because of their structural characteristics (large surface area, extensive porosity and chemical composition tunability) they are considered to have a great potential in chiral heterogeneous catalysis [63].

Enolates, enols. Shi et al. recently reported a combined chiral metal-organic framework (CDMIL-4 in Scheme 3.28) organocatalyst approach based on readily available MIL-101 grafted with (1R,2R)-1,2-diphenylethylenediamine. This has been exploited for the gram-scale, atom economic and enanthioselective synthesis of anticoagulant (S)-warfarin **38** (see Scheme 3.28) [64].

Multifunctional base–acid-metal catalysts have been used in one-pot multi-step condensation–dehydration–reduction reactions to achieve several synthetic targets in highly satisfactory yields [65]. Accordingly, jasmine-like flavor 2-cyclopentylcylopentanone **39** has been obtained from cyclopentanone in a cascade process in the presence of hydrogen and Pd/MgO as catalyst [66]. The process resulted in a minimization of the E-factor value by one order of magnitude and a

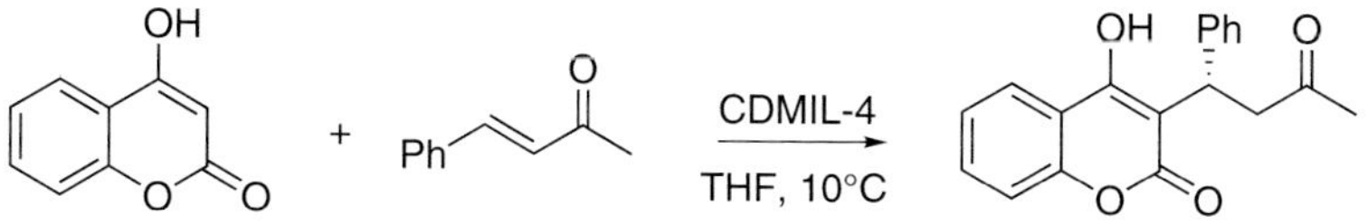

38 73% yield,
>99%ee after recrystallization

Scheme 3.28 Synthesis of (S)-warfarin [64]

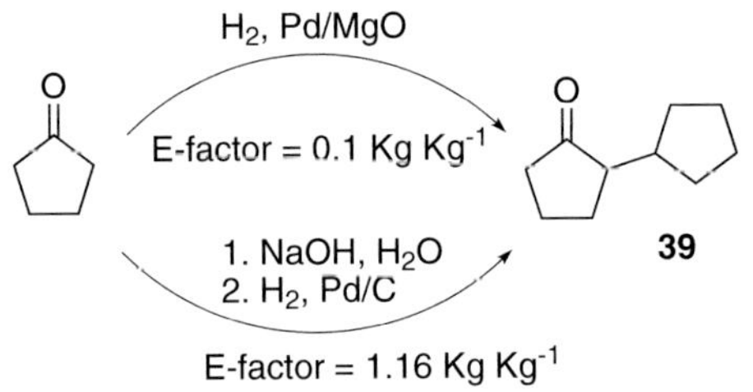

Scheme 3.29 Synthesis of jasmine-like flavor 2-cyclopentylcylopentanone **39**

more satisfying selectivity with respect to the conventional method that involved NaOH induced condensation followed by catalytic hydrogenation (Scheme 3.29).

Analogously to what observed for solid acid catalysts, commercially available solid base catalysts including basic zeolites layered-structure materials (e.g. hydrotalcite), and immobilized organic base [67]. The Sumitomo-process for the bulk production (about 2000 ton/year) of vinylbicycloheptene exploited the catalytic activity of solid superbases such as Na/NaOH/γ-Al$_2$O$_3$ for the double-bond isomerization in olefin. Thus, 5-vinylbicyclo [2.2.1] hepta-2-ene (**40**) is almost completely isomerized to 5-ethylidenebicyclo [2.2.1] hepta-2-ene (**41**), that is commercialized as a vulcanization agent. Due to the thermal instability of **41**, the side product hexahydroindene (**42**) was also observed. Anyway, under heterogeneous basic catalyst a selectivity of 99.8 % towards the desired product **41** has been observed, allowing to use a simple purification procedure (Scheme 3.30).

The combined use of different solid catalyst has been adopted for multistep procedures. The gram scale synthesis of β-nitroacrylates, an emerging class of versatile building blocks for the synthesis of highly functionalized compounds, have been optimized under heterogeneous conditions by dehydration of the corresponding nitroalkanols via in situ acid catalyzed generation of the corresponding β-nitroacetate. The reaction afforded the desired olefin **43** in >90 % yield and with E-factor values ranging from 8.8 to 15.3 kg kg^{-1} (an example in Scheme 3.31) [68].

The alkylation of methyl β-hydroxy esters **45** such as those illustrated in Scheme 3.32 has been carried out in up to 93 % yield via solvent free Mukayama condensation of an aldehyde with enolsilyl ether **44** followed by a de-silylation step, with Amberlite-Fluoride and Dowex 50Wx8-H as the elective catalysts for this process. A cyclic continuous-flow reactor operating under solvent-free conditions has been used in order to minimize the production of waste [69].

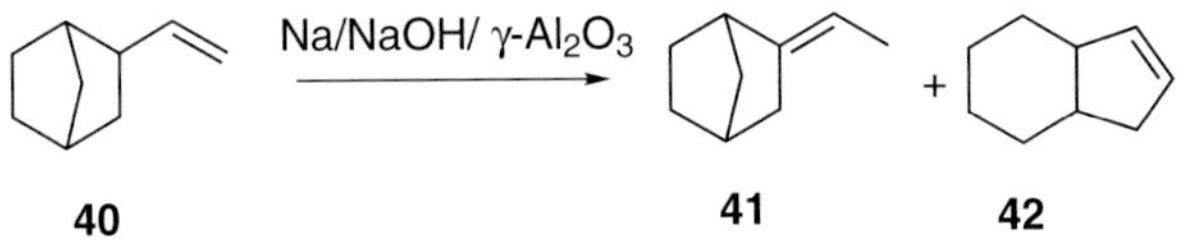

Scheme 3.30 Alkene isomerization

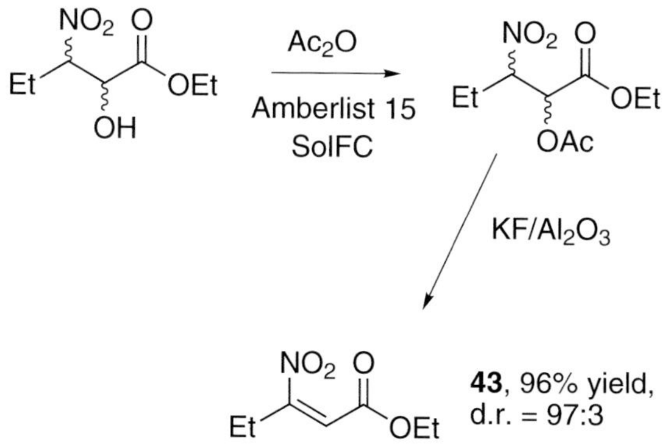

Scheme 3.31 Dehydration of a nitroalkanol

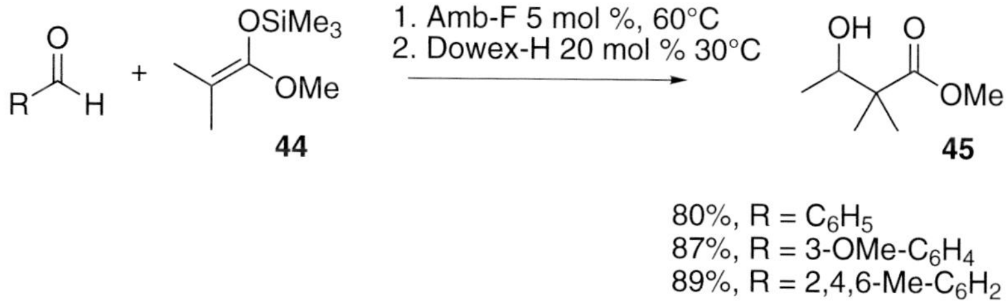

80%, R = C$_6$H$_5$
87%, R = 3-OMe-C$_6$H$_4$
89%, R = 2,4,6-Me-C$_6$H$_2$

Scheme 3.32 Solvent free Mukayama condensation [69]

Several classes of Lewis acids have found application in fine chemical production. However, the requirement of anhydrous conditions (often mandatory, since the presence of water even at impurity levels affected the reaction efficiency) made the use conventional Lewis acids such as AlCl$_3$ or FeCl$_3$ in stoichiometric amounts undesirable. Likewise, recovery and reuse of these activators is often difficult and their disposal produces a large amount of inorganic salts as waste. An appealing alternative is the use of earth metal salts such as triflates Sc(OTf)$_3$ and Yb(OTf)$_3$ that are stable in aqueous media [70] and have been used to promote different processes including esterification, cycloadditions and aromatic nitrations.

A typical example of the potentialities of these compounds is the fluoride free Mukaiyama aldol condensation involving benzaldehyde and silyl enol ether **46**. This is promoted by Yb(OTf)$_3$ in water-THF 1/4 mixture, where water is required to activate the catalyst via solvation of the Lewis acidic cation. The catalyst can be easily recovered and reused after extraction of the products (see Scheme 3.33 as an example for the synthesis of **46**) [71]. Mannich-type reactions of aldehydes, amines, and silyl enolates in water have also been performed in aqueous media and the presence of water did not affect the in situ formation of the imine.

Although less commonly, *biocatalysis* found also application in C–C bond formations. An example is the active ingredient of cholesterol-lowering drug

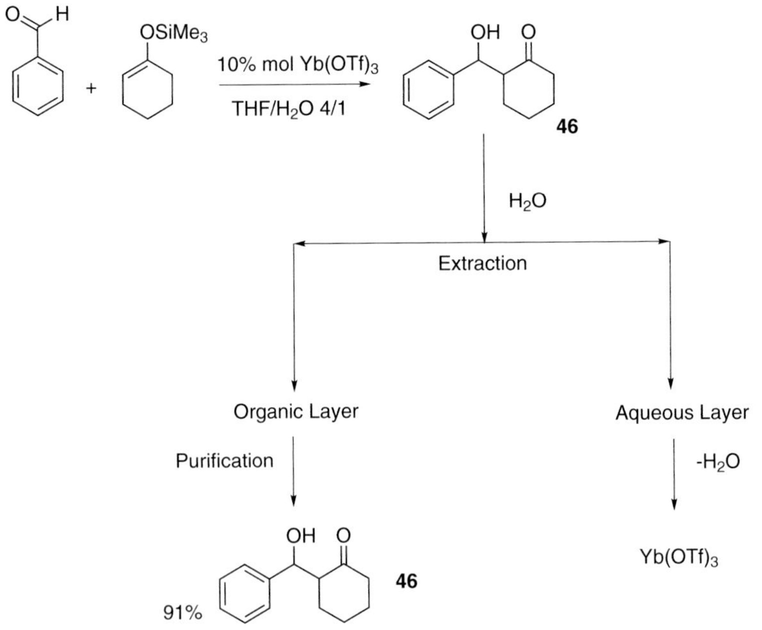

Scheme 3.33 Mukayama condensation

LipitorR (Atorvastatin calcium, **47**), which has an annual demand of ca. 100 mT. The critical step of all commercial routes to **47** is the cyanation of ethyl 3-hydroxy-4-halobutyrate **48** in alkaline solution at elevated temperature to give the corresponding hydroxynitrile **49**.

Due to the high pH value required, degradation of base sensitive substrates and products afforded several by-products, making the development of a cyanation reaction at neutral pH highly preferable. The enzymatic multistep process developed by Codexis has been successfully applied for the multiton scale production of **47**. In this process the stability and activity of enzymes has been improved by DNA shuffling technology. Thus, biocatalytic reduction of ethyl-4-chloroacetoacetate has been carried out by means of a ketoreductase (KRED) in combination with glucose and NADP-dependent glucose dehydrogenase (GDH) for cofactor regeneration. The corresponding (S)-ethyl-4-chloro-3-hydroxybutyrate **47** was isolated in 96 % yield and >99.5 % e.e.

In the second step, nucleophilic substitution of chloride by cyanide took place by using HCN at neutral pH and ambient temperature, in the presence of a halohydrindehalogenase (HHDH). The reaction was carried out in butylacetate/water mixture and the organic solvent was recycled with a 85 % efficiency at the end of the process. Despite the poor atom economy of the process (that is caused by the use of a stoichiometric amount of glucose), the use of the DNA shuffling technology for the improvement of the employed enzymes allowed for an increase of substrate loading up to 160 g l^{-1}, while the raw material was converted into

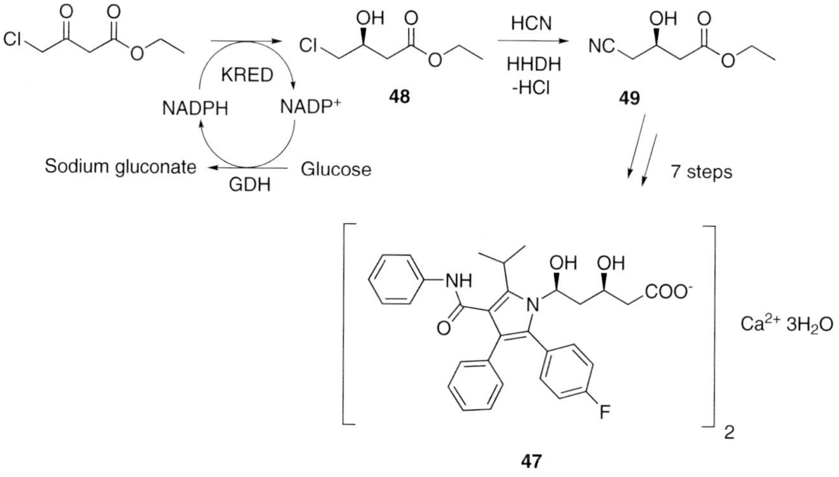

Scheme 3.34 Synthesis of Atorvastatin calcium (**47**)

products in an almost quantitative yield. Furthermore, the neutral pH avoided the presence of alkaline-induced by-products. An E-factor of 5.8 kg kg^{-1} was calculated and the main contributors were organic solvents (EtOAc and BuOAc), sodium gluconate (25 %) and inorganic salts (NaCl and Na_2SO_4). Importantly, the main waste streams are aqueous and directly biodegradable, and thus they are not considered in the E-Factor value [72] (Scheme 3.34).

A NOSE approach (compare Scheme 3.15) has been successful applied to the highly atom economic multicomponent synthesis of amidoalkyl naphthol derivatives **50** (the corresponding 1,3-amino-oxygenated moiety is largely present among a variety of natural compounds) under solvent-free reaction condition. Optimization has been obtained by having recourse to S_8 nanoparticles (prepared by annealing elemental sulfur) as catalyst (TON = 84), and takes place with a negligible production of waste (Scheme 3.35) [73].

Carbonylation. The term carbonylation covers a large variety of reaction in which carbon monoxide is incorporated into a substrate by the addition of CO to a carbon based palladium complex in the presence of various nucleophiles. The process is of crucial significance in industry.

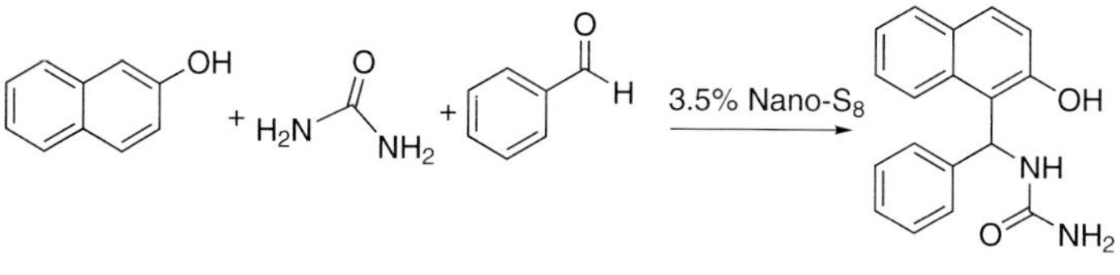

50, 98% yield, AE = 94.2%, E = 0.05 kg kg^{-1}

Scheme 3.35 NOSE approach for the synthesis of an amidoalkyl-naphthol

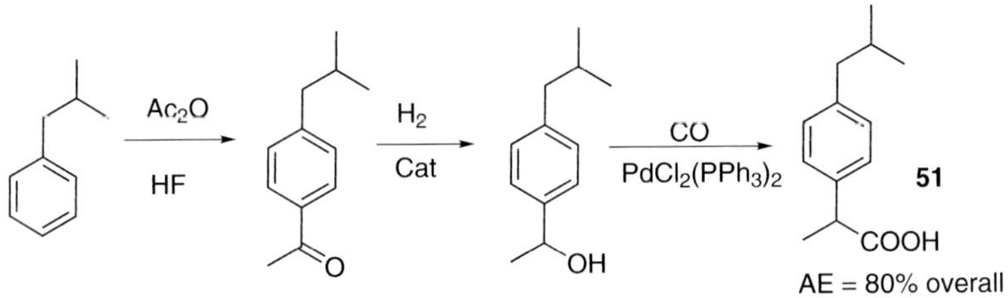

Scheme 3.36 Synthesis of Ibuprofen **51**

The traditional synthesis of α-arylpropionic acid Ibuprofen **51** involved a six non catalytic steps synthesis where less than 40 % of reactant atoms are incorporated in the final product. In contrast, the process designed by BHC (a joint venture between Hoechst Celanese Corporation and the Boots company) requires only three steps, with a calculated atom economy value of 0.8 (0.99 if by product acetic acid is recovered) (Scheme 3.22) and hazardous HF, which has the dual role of catalyst and solvent in the acylation step, is recovered and recycled with greater than 99.9 % efficiency [74] (Scheme 3.36).

As for the carbonylation step, typical conditions involved a relatively high temperature (403 K) and pressure of CO (Pco = 16.5 MPa). 1-(4-Isobutylphenyl) ethanol (IBPE) can be used as both solvent and reactant, with a quantitative conversion of the starting substrate and a selectivity towards Ibuprofen achieving 96 % [74].

In the synthesis of phenylacetic acid (**52**), the hydroxycarbonylation process efficiently replaced the previous two step cyanation of benzyl chloride followed by hydrolysis, which suffered from the formation of large amounts of salt ($1400 \, kg \, kg^{-1}$). Accordingly, the new route resulted in a 60 % reduction of inorganic waste production, avoiding, at the same time, the use of expensive and highly noxious cyanide salts (Scheme 3.37) [75].

Among the industrial processes involving carbonylation as key-step, rhodium or cobalt complexes catalyzed hydroformylation is probably the most significant one [76] and the main application is the synthesis of butyrraldehyde **53** from propylene and potentially renewable derived syngas (Scheme 3.38). The final product is easily separated from the catalyst by simple distillation of the crude product. However, when the reaction was carried out under homogeneous conditions, the presence of some oligomers produced during the process resulted in the partial deactivation of

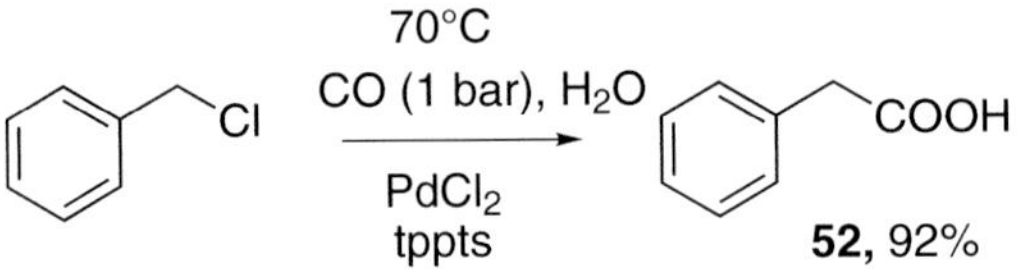

Scheme 3.37 Synthesis of phenyl acetic acid **52** [75]

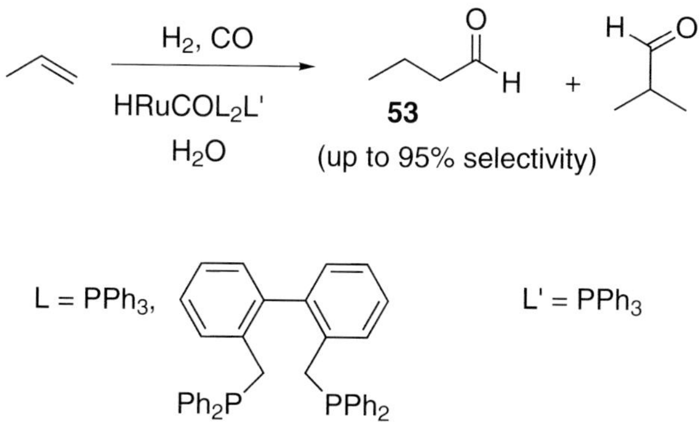

Scheme 3.38 Hydroformylation of propilene to butirraldehyde **53**

the catalyst. Another issue was the application of this process to the synthesis of aldehydes heavier than butyrraldehyde, the purification of which is still a challenge.

At this regard, aqueous-organic biphasic catalysis offers an elegant approach to overcome the above cited limitations. This has been applied for the hydroformylation of lower olefins (propylene to pentene) and performed in two locations, Ruhrchemie/Oberhausen, Germany, and Yeochun, Korea, for a total aldehyde amount of about 800,000 ton year^{-1} (mainly C_4 and C_5 aldehydes). The process is based on the ability to bring the rhodium based catalyst for the hydroformylation into an aqueous phase. In this case, water is used as a "mobile-immobilizing" phase, while non-polar products are present as a water non miscible organic phase and can be simply removed by extraction or decantation. Historically, the ligand mostly used in industry was tris(meta-sulfonated) triphenylphosphine (TPPS), leading to the formation of a water soluble (ca. 1.1 kg l^{-1}) catalyst [HRh(CO)-(m-SO$_3$NaC$_6$H$_4$)$_3$P}$_3$] bearing nine sulfonate substituents (i.e., three per P atom) and accordingly highly soluble in water. However, the efficiency of the hydroformylation of higher olefins is poor, because of the limited solubility of these substrates in the aqueous phase, which results in rates too slow for commercial applications (see Scheme 3.38) [77].

In an ideal system, synthesis and separation of the products should take place under the same conditions, in order to keep the catalyst in its active state and, at the same time, to optimize the reactor engineering (Scheme 3.39). Fluorous biphasic systems show a temperature dependent phase behavior because a single phase exists at the reaction temperature allowing good reaction rates, but the phases separate on cooling. In this case, the key issue is the design of a catalyst that is efficient in hydroformylation processes, while remaining confined in the perfluorinated media [78]. Perperi et al. proposed the hydroformilation of 1-octene in perfluoromethyl-cyclohexane in the presence of a fluorophilic rhodium based complex (Scheme 3.40). The reaction was carried out in a reactor from which the aldehyde **54** generated was continuously removed, while the catalyst phase was recycled to the reactor [79]. Under these conditions, the catalyst exhibited a turn over frequency (TOF) value

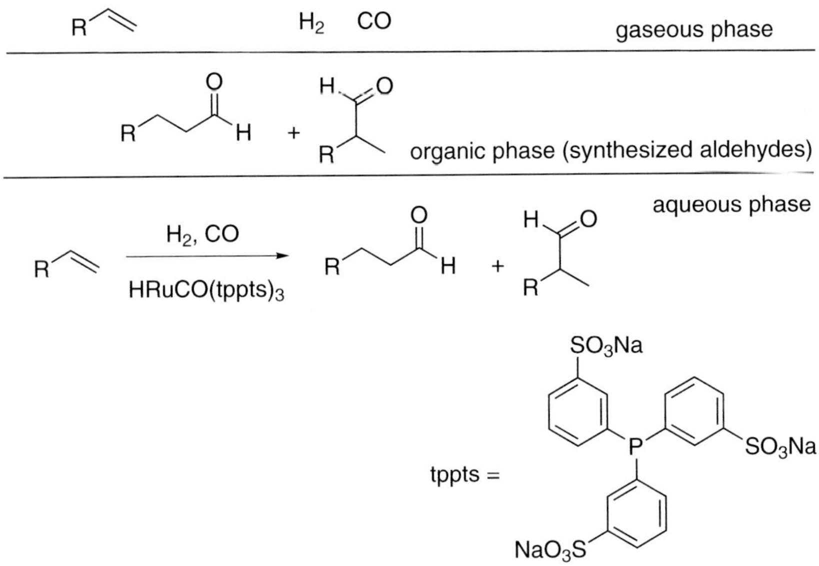

Scheme 3.39 Aqueous/organic phase hydroformylation

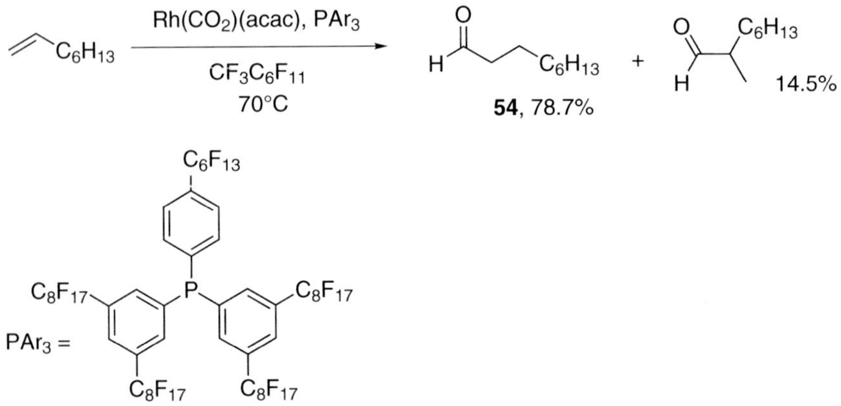

Scheme 3.40 Fluorous phase hydroformylation of 1-octene

comparable with that observed with commercial homogeneous rhodium catalysts used for propylene hydroformylation (TOF = 750 h^{-1}).

An analogous rhodium based catalyst has been exploited in the hydroformylation of alkenes in supercritical carbon dioxide. When using PEt$_3$ as ligand, the process showed efficiencies analogous to those observed when using toluene as solvent, with a up to 97 % selectivity [80, 81]. The use of imidazolium based ionic liquids as immobilizing phase has been also successfully implemented [82].

Further, Webb et al. reported an interesting process for the hydroformylation of relatively low volatility alkenes in a continuous flow system. In the developed system, thanks to the presence of imidazolium salts of the sulfinate phosphine ligands, the catalyst was confined in the ionic liquid (best results were obtained with 1-alkyl-3-methylimidazolium bis(trifluoromethanesulfonyl)amides) while the liquid alkene and gaseous reagents were dissolved in supercritical CO_2, which acted both as solvent and as transport vector for the synthesized aldehydes. Decompression of the reaction mixture yielded the desired products, avoiding any rhodium leaching into the product stream [83].

Synthesis of heterocycles. Multicomponent reactions often represent the ultimate choice to synthesize heterocycles in eco-sustainable, efficient processes. Microwave activation is often used. This is the case of the acid-catalyzed multicomponent synthesis of substituted quinolines reported by Kulkarni and Torok [84]. The desired quinolines **24** are obtained through a multicomponent domino reaction from anilines, aldehydes and terminal aryl alkynes, in the presence of environmentally benign montmorillonite K-10 as acid catalyst. Interestingly, a nearly 90 % atom economy was calculated, and the use of microwave activation strongly decreased the reaction time (see some examples in Scheme 3.41).

When carried out on smaller scale (down to 10 mmol), the use of microwave irradiation generally resulted in an improvement of the green metrics of the process, although the values obtained were still far from satisfactory. As an example, the microwave induced solvent free synthesis of pyrazoles **56** proposed by Corradi et al. results in an impressive improvement with respect to the classical protocol that involved the use of a significant amount of methanol as solvents. The calculated parameters, both mass-related (E-Factor and PMI) and environment-related (E_{IN}, E_{OUT}) obtained by means of the EATOS software were reduced by about 3 order magnitude, though still remaining significant (E_{IN} up to ca. 170 PEI kg^{-1}), with the largest contribution given by the auxiliaries required for the isolation of **25**, viz. extraction with diethyl ether and recrystallization of the crude product from ethanol or cyclohexane (Scheme 3.42) [85].

The contribution of the work-up/isolation in the synthesis of *1-H*-pyrazole derivatives has been also pointed out by Martins et al. In this case, the solvent free microwave induced synthesis of pentafluorophenylsubstituted compounds **57** resulted to have the most favorable reaction yield and RME value. However, the

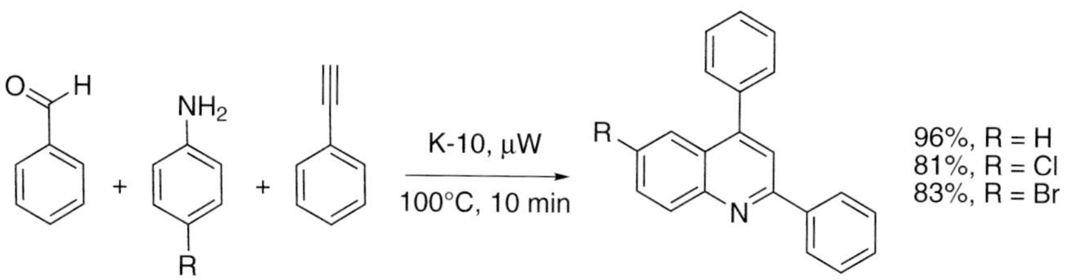

96%, R = H
81%, R = Cl
83%, R = Br

55

Scheme 3.41 Three component synthesis of quinolines

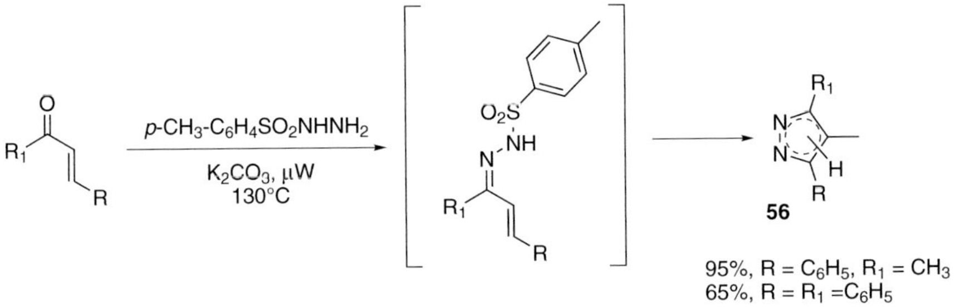

Scheme 3.42 Synthesis of pyrazoles

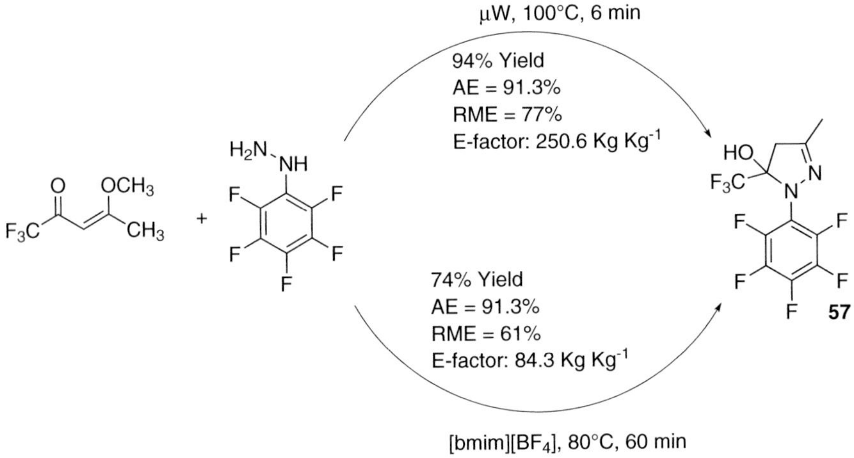

Scheme 3.43 Synthesis of pentafluorophenylsubstituted 1-H-pyrazoline **57** [86]

more elaborate work up procedure and the huge amount of solvent used in the purification of the final compound strongly increased the generation of waste, and the resulting E-Factor was 3 times higher than that measured for the analogous conventional thermal approach (Scheme 3.43) [86].

Henriques et al. recently reported the multicomponent synthesis of porphyrin derivative **58** in neat water as the solvent (see Scheme 3.44), which occurred under microwave irradiation at 473 K and pressure above 16 bar. A low environmental impact of the reaction was demonstrated by the lowest E-Factor (35 kg kg^{-1}) value and highest Ecoscale score (50.5) ever reported for the synthesis of these compounds [87].

Cycloadditions. The elective way to synthesize cyclobutanes and cyclobutenes is the photochemically induced [2 + 2] cycloaddition. Booker–Milburn optimized the cycloaddition between maleimide **59** and 1-hexyne to afford cyclobutene **60**. The reaction has been performed under flow conditions, in a continuous flow

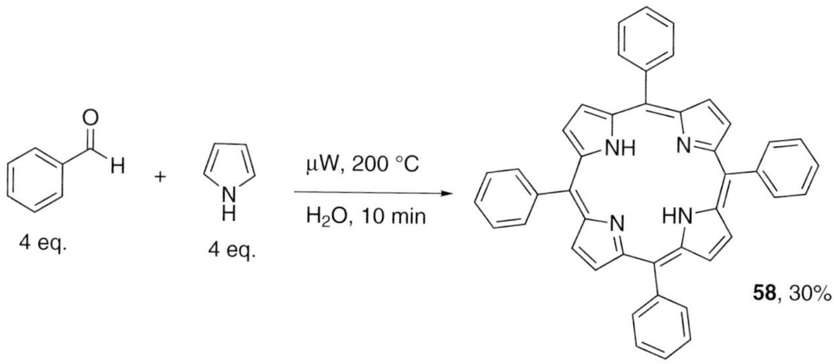

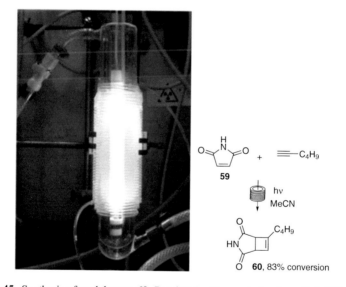

Scheme 3.44 Microwave assisted multicomponent synthesis of porphyrins [87]

Scheme 3.45 Synthesis of cyclobutene **60**. Reprinted with permission from Ref. [88]

photochemical system composed by a Fluorinated Ethylene Propylene (FEP) tubing coiled three times around a water cooled medium pressure Hg lamp. The reaction gave the desired product in >90 % yield, with a productivity of nearly 0.7 kg day^{-1} (Scheme 3.45). Both electrical energy for lamp operation and cooling water contributed to energy balance [88].

Yoon et al. developed a visible light photoredox (ruthenium catalyzed) [2 + 2] cyclization of enones and styrenes that was carried out on a gram scale and under sunlight irradiation (an example in Scheme 3.46) [89].

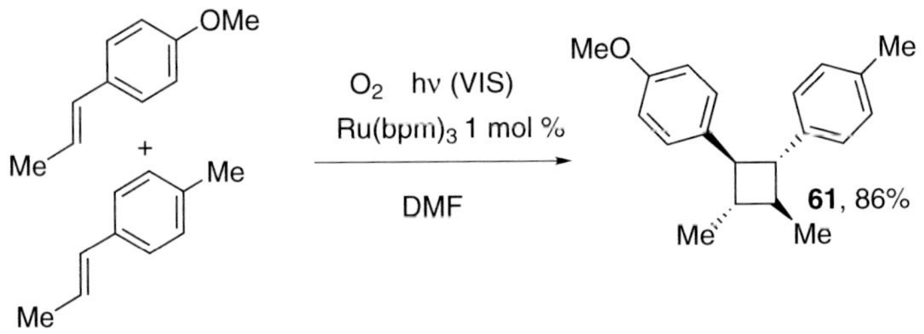

Scheme 3.46 Photocatalyzed 2+2 cycloaddition. Ref. [89b]

References

1. Albini A, Fagnoni M (2008) 1908 Giacomo Ciamician and the concept of green chemistry. ChemSusChem 1:63–66
2. Lancaster M (2010) Green chemistry: an introductory text. Royal Chemical Society, London
3. Hoelderich WF (2000) Environmentally benign manufacturing of fine and intermediate chemicals. Cat Today 62:115–130
4. For review on the contribution of catalysis to green chemistry see: (a) Anastas PT, Kirchhoff MM, Williamson TC (2001) Catalysis as a foundational pillar of green chemistry. Appl Cat A 221:3–13. (b) Anastas PT, Bartlett LB, Kirchhoff MM, Williamson TC (2000) The role of catalysis in the design, development, and implementation of green chemistry. Cat Today 55:11–22. Sheldon R, Arends IWCE, Hanefeld U (2007) Green Chemistry and Catalysis. Wiley VCH, Germany
5. Sheldon RA (1997) Catalysis: the key to waste minimization. J Chem Tech Biotechnol 68:381–388
6. Sheldon RA (1999) Downing heterogeneous catalytic transformations for environmentally friendly production. Appl Catal A 189:163–183
7. See for reviews Climent MJ, Corma A, Iborra S (2011) Heterogeneous catalysts for the one-pot synthesis of chemicals and fine chemicals. Chem Rev 111:1072–1133. Hattori, H (2001) Solid base catalysts: generation of basic sites and application to organic synthesis. Appl Cat A: General 222:247–259
8. Pollet P, Hart RJ, Eckert C A, Liotta CL (2010) Organic aqueous tunable solvents (OATS): a vehicle for coupling reactions and separations. Acc Chem Res 43:1237–1245
9. Vincent JM (2012) Fluorous catalysis: from the origin to recent advances. Topics Curr Chem 308:153–174
10. Dyson PJ (2004) Biphasic synthesis. In McCleverty JA, Meyer TJ (eds) Comprehensive coordination chemistry II, vol 1, pp 689–695
11. Nelson A (1999) Asymmetric phase-transfer catalysis. Angew Chem Int Ed 38:1583–1585. Shirakawa S, Maruoka K (2013) Recent developments in asymmetric phase-transfer reactions. Angew Chem Int Ed 52:4312–4348
12. Peach J, Eastoe J (2014) Supercritical carbon dioxide: a solvent like no other. Beilstein J Org Chem 10:1878–1895
13. Jindal R, Sablok A (2015) Preparation and applications of room temperature ionic liquids in organic synthesis: a review on recent efforts. Curr Green Chem 2:135–155
14. Ran N, Zhao L, Chen Z, Tao J (2008) Recent applications of biocatalysis in developing green chemistry for chemical synthesis at the industrial scale. Green Chem 10:361–372. Gupta P, Mahajan A (2015) Green chemistry approaches as sustainable alternatives to conventional strategies in the pharmaceutical industry. RSC Adv 5:26686–26705

15. See for reviews Wohlgemuth R (2010) Biocatalysis-key to sustainable industrial chemistry. Curr Opin Biotech 21:713–724; Tao J, Xu J-H (2009) Biocatalysis in development of green pharmaceutical processes. Curr Opin Chem Biol 13:43–50. Ran N, Zhao L (eds) (2011) Biocatalysis for green chemistry and chemical processes development. Wiley, Hoboken, New Jersey. Patel RN (2013) Biocatalytic synthesis of chiral alcohols and amino acids for development of pharmaceuticals. Biomolecules 3:741–777

16. (a) Nüchter M, Müller U, Ondruschka B, Tied A, Lautenschläger W (2003) Microwave assisted synthesis. A critical technology overview. Chem Eng Technol 26:1207–1216 (b) Lidström P, Tieney J, Wathey B, Westmann J (2001) Microwave organic synthesis. A review. Tetrahedron 57:9225–9283. (c) Polshettiwar V, Varma RS (2008) Aqueous microwave chemistry: a clean and green synthetic tool for rapid drug discovery. Chem Soc Rev 37:1546–1557. (d) Gronnow MJ, White RJ, Clark JH, Macquarrie DJ (2005) Energy efficiency in chemical reactions: a comparative study of different reaction techniques. Org Proc Res Dev 9:516–518. (e) Roberts BA, Strauss CR (2005) Toward rapid, "green", predictable microwave-assisted synthesis. Acc Chem Res 38:653–661

17. Fujishima A, Rao TN, Tryk DA (2000) Titanium dioxide photocatalysis. J Photochem Photobiol C: Photochem Rev 1:1–21. Protti S, Albini A Serpone N (2014) Photocatalytic generation of solar fuels from the reduction of H_2O and CO_2: a look at the patent literature. Phys Chem Chem Phys 16:19790–19827

18. Ravelli D, Dondi D, Fagnoni M, Albini A (2009) Photocatalysis. A multi-faceted concept for green chemistry. Chem Soc Rev 38:1999–2011

19. Sato K, Aoki M, Noyori R (1998) A "green" route to adipic acid: direct oxidation of cyclohexenes with 30 percent hydrogen peroxide. Science 281:1646–1647

20. Noyori R, Aoki M, Sato, K (2003) Green oxidation with aqueous hydrogen peroxide. Chem Commun 16:1977–1986

21. Kinen CO, Rossi LI, de Rossi RH (2009) The development of an environmentally benign sulfide oxidation procedure and its assessment by green chemistry metrics. Green Chem 11:223–228

22. Taramasso M, Perego G, Notari B (1984) Preparation of porous crystalline synthetic material comprised of silicon and titanium oxides. US Patent 4410501 A

23. Thiele GF, Roland E (1997) Propylene epoxidation with hydrogen peroxide and titanium silicalite catalyst: activity, deactivation and regeneration of the catalyst. J Mol Cat A: Chem 117:351–356

24. Schenck GO, Ziegler (1944) Die Synthese des Ascaridols. Naturwissenschaft 32:157

25. Schenk GO, Ohloff G, Klein E (1968) Mixtures of oxygenated acyclic terpenes. US Patent 3,382,276

26. Ravelli D, Protti S, Neri P, Fagnoni M, Albini A (2011) Photochemical technologies assessed: the case of rose oxide. Green Chem 13:1876–1884

27. Varma RS (1997) The presidential green chemistry challenge awards program, summary of 1997 award entries and recipients, EPA744-S-97-001. U.S. Environmental Protection Agency, Office of Pollution Prevention and Toxics, Washington, DC, p 9

28. Pharmacia & Upjohn, An alternative synthesis of bisnoraldehyde, an intermediate to progesterone and corticosteroids. In: The presidential green chemistry challenge awards program summary of 1996 award. Entries and recipients. http://www2.epa.gov/sites/production/files/documents/award_entries_and_recipients1996.pdf. Last access 14 Jan 2015

29. Haggiage E, Coyle EE, Joyce K, Oelgemoeller M (2009) Green photochemistry: solar chemical synthesis of 5-amido-1,4-naphthoquinones. Green Chem 11:318–321

30. Le Bars J, Dakka J, Sheldon RA (1996) Ammoximation of cyclohexanone and hydroxyaromatic ketones over titanium molecular sieves. Appl Catal A: General 136:69–80

31. Sheldon RA (2000) Atom efficiency and catalysis in organic synthesis. Pure Appl Chem 72:1233–1246

32. Stewart JD, Reed KW, Martinez CA, Zhu J, Chen G, Kayser MM (1998) Recombinant baker's yeast as a whole-cell catalyst for asymmetric Baeyer–Villiger oxidations. J Am Chem Soc 120:3541–3547

33. Kohlmann C, Leuchs S, Greiner L, Leitner W (2011) Continuous biocatalytic synthesis of (R)-2-octanol with integrated product separation. Green Chem 13:1430–1436
34. Wu X, Wang J, Chen C, Liu N, Chen Y (2009) Enantioselective synthesis of ethyl (S)-2-hydroxy-4-phenylbutyrate by recombinant diketoreductase. Tetrahedron Asymmetry 20:2504–2509
35. Easwar S, Argade NP (2003) Amano PS-catalysed enantioselective acylation of (±)-α-methyl-1,3-benzodioxole-5-ethanol: an efficient resolution of chiral intermediates of the remarkable antiepileptic drug candidate, (−)-talampanel Tetrahedron Asym 14:333–337 (c) Nakamura K, Yamanaka R, Matsuda T, Harada T (2003) Recent developments in asymmetric reduction of ketones with biocatalysts. Tetrahedron Asymm 14:2659–2681
36. Savile CK, Janey JM, Mundorff EC, Moore JC, Tam S, Jarvis WR, Colbeck JC, Krebber A, Fleitz FJ, Brands J, Devine PN, Huisman GW, Hughes GJ (2010) Biocatalytic asymmetric synthesis of chiral amines from ketones applied to sitagliptin manufacture. Science 329:305–309
37. Hansen KB, Hsiao Y, Xu F, Rivera N, Clausen A, Kubryk M, Krska S, Rosner T, Simmons B, Balsells J, Ikemoto N, Sun Y, Spindler F, Malan C, Grabowski EJJ, Armstrong JD III (2009) Highly efficient asymmetric synthesis of sitagliptin. J Am Chem Soc 131:8796–8804
38. (a) Adams JP, Alder CM, Andrews I, Bullion AM, Campbell-Crawford M, Darcy MG, Hayler JD, Henderson RK, Oare CA, Pendrak I, Redman AM, Shuster LE, Sneddon HF, Walker MD (2013) Development of GSK's reagent guides—embedding sustainability into reagent selection. Green Chem 15:1542–1549. An analogous scale has been compiled for the choice of solvents (see Fig 2 in Ch 5). (b) Clark JH (2002) Solid acids for green chemistry. Acc Chem Res 35:791–797
39. Comerford JW, Clark JH, Macquarrie DJ, Breeden SW (2009) Clean, reusable and low cost heterogeneous catalyst for amide synthesis. Chem Comm 18:2562–2564
40. Caldwell N, Jamieson C, Simpson I, Watson AJB (2013) Development of a sustainable catalytic ester amidation process. ACS Sustain Chem Eng 1:1339–1344
41. Das VK, Devib RR, Thakur AJ (2013) Recyclable, highly efficient and low cost nano-MgO for amide synthesis under SFRC: a convenient and greener 'NOSE' approach. Appl Cat A: Gen 456:118–125
42. Verweij J, de Vroom E (1993) Industrial transformations of penicillins and cephalosporins. Recl Trav Chim Pays-Bas 112:66–81
43. Bruggink A, Roos EC, de Vroom E (1998) Penicillin acylase in the industrial production of β-lactam antibiotics. Org Proc Res Dev 2:128–133
44. Arroyo M, de la Mata I, Acebal C, Castillón MP (2003) Biotechnological applications of penicillin acylases: state-of-the-art. Appl Microbiol Biotechnol 60:507–514. Youshko MI, van Langen LM, de Vroom E, van Rantwijk F, Sheldon RA, Svedas, VK (2001) Highly efficient synthesis of ampicillin in an "Aqueous Solution-Precipitate" system: repetitive addition of substrates in a semicontinuous process. Biotech Bioengin 73:426–430. Wegman MA, Janssen MHA, van Rantwijk F, Sheldon RA (2001) Towards biocatalytic synthesis of ß-lactam antibiotics. Adv Synth Catal 343:559–576
45. Hölderich WH, Dahlhoff G, Ichihashi H, Sugita K (2003) Method for producing ε-caprolactam and reactor for the method. US Patent 6531595 B2
46. Barnard TM, Vanier GS, Collins MJ Jr (2006) Scale-up of the green synthesis of azacycloalkanes and isoindolines under microwave irradiation. Org Proc Res Dev 10:1233–1237
47. Ryu I, Tani A, Fukuyama T, Ravelli R, Montanaro S, Fagnoni (2013) Efficient C–H/C–N and C–H/C–CO–N conversion via decatungstate-photoinduced alkylation of diisopropyl azodicarboxylate. Org Lett 15:2554–2557
48. Paravidino M, Hanefeld U (2011) Enzymatic acylation: assessing the greenness of different acyl donors. Green Chem 13:2651–2657
49. Cabrera Z, Fernandez-Lorente G, Fernandez-Lafuente R, Palomo JM, Guisan JM (2009) Enhancement of Novozym-435 catalytic properties by physical or chemical modification. Process Biochem 44:226–231
50. Korupp C, Weberskirch R, Muller JJ, Liese A, Hilterhaus L (2010) Scale-up of lipase-catalyzed polyester synthesis. Org Process Res Dev 14:1118–1124

51. Andraos JA (2012) Green metrics assessment of phosgene and phosgene-free syntheses of industrially important commodity chemicals. Pure Appl Chem 84:827–860
52. Tundo P, Selva M, Marques CA (1996). In: Anastas PT, Williamson TC (eds) Green chemistry: designing chemistry for the environment, Ch. 7. American Chemical Society, Washington, DC, p 81
53. Cooke M, Clark J, Breeden S (2009) Lewis acid catalysed microwave-assisted synthesis of diaryl sulfones and comparison of associated carbon dioxide emissions. J Mol Catalysis A: Chem 103:132–136
54. Armor JN (1992) Environmental catalysis. Appl Catal B: Environ 1:221–256
55. Davis ME (1993) New vistas in zeolite and molecular sieve catalysis. Acc Chem Res 26:111–115
56. See for example Fong YY, Abdullah AZ, Ahmad AL, Bhatia S (2008) Development of functionalized zeolite membrane and its potential role as reactor combined separator for para-xylene production from xylene isomers. Chem Eng J 139:172–193
57. Hoefnagel AJ, van Bekkum H (1993) Direct Fries reaction of resorcinol with benzoic acids catalyzed by zeolite H-beta. Appl Catal A: Gen 97:87–102
58. Yadav GD, Kamble SB (2009) Synthesis of carvacrol by Friedel–Crafts alkylation of o-cresol with isopropanol using superacidic catalyst UDCaT-5. J Chem Technol Biotechnol 84:1499–1508
59. Yadav GD, Kamble SB (2012) Atom efficient Friedel–Crafts acylation of toluene with propionic anhydride over solid mesoporous superacid UDCaT-5. Appl Cat A: Gen 433–434:265–274
60. Ishihara K, Kubota M, Kurihara H, Yamamoto H (1996) Scandium trifluoromethanesulfonate as an extremely active Lewis acid catalyst in acylation of alcohols with acid anhydrides and mixed anhydrides. J Org Chem 61:4560–4567
61. Korea R, Srivastava R, Satpatib B (2015) Synthesis of industrially important aromatic and heterocyclic ketones using hierarchical ZSM-5 and beta zeolites. Appl Catal A: Gen 493:129–141
62. Chaube VD, Moreau P, Finiels A, Ramaswamy AV, Singh AP (2002) A novel single step selective synthesis of 4-hydroxybenzophenone (4-HBP) using zeolite H-beta. Cat Lett 79:89–94
63. See for reviews: Farrusseng D, Aguado S, Pinel C (2009) Metal-organic frameworks: opportunities for catalysis. Angew Chem Int Ed 48:7502–7513. Lee JY, Farha OK, Roberts J, Scheidt KA, Nguyen ST, Hupp JT (2009) Metal–organic framework materials as catalysts. Chem Soc Rev 38:1450–1459
64. Shi T, Guo Z, Yu H, Xie J, Zhong Y, Zhua W (2013) Atom-economic synthesis of optically active warfarin anticoagulant over a chiral MOF organocatalyst. Adv Synth Cat 355:2538–2543
65. Climent MJ, Corma A, Iborra S, Mifsud M, Velty A (2010) New one-pot multistep process with multifunctional catalysts: decreasing the E-factor in the synthesis of fine chemicals. Green Chem 12:99–107
66. For other examples related to the use of multifunctional catalysts see Climent MJ, Corma A, Iborra S, Sabater MJ (2014) Heterogeneous catalysis for tandem reactions. ACS Catal 4:870–891
67. Tanabea K, Hoelderich WF (1999) Industrial application of solid acid-base catalysts. Appl Catal A: Gen 181:399–434
68. Palmieri A, Gabrielli S, Ballini R (2013) Low impact synthesis of β-nitroacrylates under fully heterogeneous conditions. Green Chem 15:2344–2348
69. Fringuelli D, Lanari D, Pizzo F, Vaccaro L (2010) An E-factor minimized protocol for the preparation of methyl β-hydroxy esters. Green Chem 12:1301–1305
70. Kobayashi S, Manabe K (2000) Green Lewis acid catalysis in organic synthesis. Pure Appl Chem 72:1373–1380. Kobayashi S, Manabe K (2002) Development of novel Lewis acid catalysts for selective organic reactions in aqueous media. Acc Chem Res 35:209–217
71. Kobayashi S, Hachiya I (1994) Lanthanide triflates as water-tolerant Lewis acids activation of commercial formaldehyde solution and use in the aldol reaction of silyl enol ethers with aldehydes in aqueous media. J Org Chem 59:3590–3596

72. Fox RJ, Davis SC, Mundorff EC, Newman LM, Gavrilovic V, Ma SK, Chung LM, Ching L, Tam S, Muley S, Grate J, Gruber J, Whitman JC, Sheldon RA, Huisman GW (2007) Improving catalytic function by ProSAR-driven enzyme evolution. Nat Biotech 25:338–344
73. Das VK, Borah M, Thakur AJ (2013) Piper-betle-shaped nano-S-catalyzed synthesis of 1-amidoalkyl-2-naphthols under solvent-free reaction condition: a greener "nanoparticle-catalyzed organic synthesis enhancement" approach. J Org Chem 78:3361–3366
74. Hendricks JD, Mott GN (1992) Method for producing ibuprofen. Hoechst Celansese Corporation, US Patent 5 166 418. Process for the carbonylation of 1-(4-isobutylphenyl) ethanol in the presence of ibuprofen. Eur Patent Appl, EP 460 905, 1991, Chem Abstr, 116 (1992) 83378
75. Qiu Z, He Y, Zheng D, Liu F (2005) Study on the synthesis of phenylacetic acid by carbonylation of benzyl chloride under normal pressure. J Nat Gas Chem 14:40–46. See also Cornils D, Herrmann WA (eds) (2006) Aqueous-phase organometallic catalysis: concepts and applications. Wiley VCH, Germany
76. Evans D, Osborn JA, Wilkinson G (1968) Hydroformylation of alkenes by use of rhodium complex catalysts. J Chem Soc A 3133–3142
77. Bohnen HW, Cornils B (2002) Hydroformylation of alkenes: an industrial view of the status and importance. Adv Catal 47:1–64. Cornils B (1998) Industrial aqueous biphasic catalysis: status and directions. Org Proc Res Dev 2:121–127
78. Horvàth I (1998) Fluorous biphase chemistry. Acc Chem Res 31:641–650
79. Perperi E, Huang Y, Angeli P, Manos C, Mathison CR, Cole-Hamilton DJ, Adams DJ, Hope EG (2004) The design of a continuous reactor for fluorous biphasic reactions under pressure and its use in alkene hydroformylation. Dalton Trans 14:2062–2064. Adams DJ, Bennett JA, Cole-Hamilton DJ, Hope DJ, Hopewell J, Kight J, Pogorzelec P, Stuart AM (2005) Rhodium catalyzed hydroformylation of alkenes using highly fluorophilic phosphines. Dalton Trans 24:3862–3867. Bach I, Cole-Hamilton DJ (1998) Hydroformylation of hex-1-ene in supercritical carbon dioxide catalyzed by rhodium trialkylphosphine complexes. Chem Commun 14:1463–1464
80. Webb PW, Kunene TE, Cole-Hamilton DJ (2005) Continuous flow homogeneous hydroformylation of alkenes using supercritical fluids. Green Chem 7:373–379
81. Sellin MF, Bach I, Webster JM, Montilla F, Rosa V, Avilés T, Poliakoff M, Cole-Hamilton DJ (2002) Hydroformylation of alkenes in supercritical carbon dioxide catalysed by rhodium trialkylphosphine complexes. J Chem Soc Dalton Trans 4569–4576
82. Chauvin Y, Mussmann L, Olivier H (1995) A novel class of versatile solvents for two-phase catalysis: hydrogenation, isomerization, and hydroformylation of alkenes catalyzed by rhodium complexes in liquid 1,3-dialkylimidazolium salts. Angew Chem Int Ed Eng 34:2698–2700
83. Webb PB, Sellin MF, Kunene TE, Williamson S, Slawin AMZ, Cole-Hamilton DJ (2003) Continuous flow hydroformylation of alkenes in supercritical fluid-ionic liquid biphasic systems. J Am Chem Soc 125:15577–15588
84. Kulkarni A, Torok B (2010) Microwave-assisted multicomponent domino cyclization–aromatization: an efficient approach for the synthesis of substituted quinolones. Green Chem 12:875–878
85. Corradi A, Leonelli C, Rizzuti A, Rosa R, Veronesi P, Grandi R, Baldassari S, Villa C (2007) New "green" approaches to the synthesis of pyrazole derivatives. Molecules 12:1482–1495
86. Martins MAP, Beck PH, Buriol L, Frizzo CP, Moreira DN, Marzari MRB, Zanatta M, Machado P, Zanatta N, Bonacorso HG (2013) Evaluation of the synthesis of 1-(pentafluorophenyl)-4,5-dihydro-1H-pyrazoles using green metrics. Monatsh Chem 144:1043–1050
87. Henriques CA, Pinto SMA, Aquino GLB, Pineiro M, Calvete MJF, Pereira MM (2014) Ecofriendly porphyrin synthesis by using water under microwave irradiation. Chem Sus Chem 7:2821–2824

88. (a) Hook BDA, Dohle WP, Hirst R, Pickworth M, Berry MB, Booker–Milburn KI (2005) A practical flow reactor for continuous organic photochemistry. J Org Chem 70:7558–7564. (b) Knowles JP, Elliott LD, Booker–Milburn KI (2012) Flow photochemistry: old light through new windows. Beilstein J Org Chem 8:2025–2052

89. (a) Yoon TP, Ischay M, Du J (2010) Visible light photocatalysis as a greener approach to photochemical synthesis. Nat Chem 2:527–532. (b) Ischay MA, Ament MS, Yoon TP (2012) Crossed intermolecular [2 + 2] cycloaddition of styrenes by visible light photocatalysis. Chem Sci 3:2807–2811

Chapter 4
Renewable Resources: From Refinery to Bio-refinery

Abstract Chemists are educated to consider petrochemicals as the source of both new molecules and energy. However, biological material (biomass) from living or recently living organisms, not metabolized for thousands of years into petrol and coal, offers an alternative feedstock that is elaborated in the so called bio-refineries to a variety of platform chemicals (alcohols, acids, esters, carbonyls, hydrocarbons). The environmental performance of fermentative and biocatalytic methods is compared with that from fossil fuel.

Keywords Renewable feedstock · Biorefinery · Platform chemicals

4.1 Shifting to Renewable

The use of non renewable fossil as fuels and source of chemicals is at the basis of modern chemistry and the enormous development of modern society, but it also led to a kind of psychological dependence of chemist towards this resource. As stated by Metzger in 2004 [1], "chemists learn from the very beginning to think in terms of *petrochemical product* lines." However, whereas they "continue to prefer simple petrochemical molecules as feedstock to develop catalysts and reactions, the growing trend among the newly forming green chemistry community is to develop green catalysts, green solvents, and green reactions around *renewable source materials*". This is in accordance with the seventh principle of Green Chemistry, "A raw material or feedstock should be renewable rather than depleting whenever technically and economically practicable." By incorporating the above statement by Metzger [1], one can say that "The use of *renewable raw materials or feedstock* rather than depleting fossil must be made technically and economically practicable." Fuel (accounting for the largest amount), commodities and fine chemicals are to be obtained from a cheap, renewable feedstock alternative to fossil fuels, a change of perspective that involves huge amounts of materials. What is required is a source of carbon analogous to petrol, where a complex mixture of chemicals can be

© The Author(s) 2016
A. Albini and S. Protti, *Paradigms in Green Chemistry and Technology*,
SpringerBriefs in Green Chemistry for Sustainability,
DOI 10.1007/978-3-319-25895-9_4

converted in several high-value molecules, through the development of efficient refinement procedure including cracking, chemical modification and separation in the so-called bio-refineries (see below for a detailed definition). The obvious alternative to fossils is biological material (*biomass*) from living or recently living organisms, not metabolized for thousands of years into petrol and coal, thus not concurring to CO_2 emissions. Sunlight is the obvious source of energy for biomass growth, and the huge amount of energy (4.3×10^{20} J h^{-1}) coming from the sun [2] allows for the production of ca. 170 billion tons of biomass per year (more than 10^{10} ton of carbon per year [3]). 75 % of the whole amount belongs to the class of carbohydrates, which are considered one of the most important chemical feedstock for the future [4] and the remaining part is formed by lignin (20 %) and other natural products (5 %, e.g. oils and fats, proteins and nucleic acids). Despite the enormous amount of energy available, it has been calculated that less than 5 % of such products is used for human purposes (both food and non food). As for carbohydrates, the preferred non-food applications of lignocellulosic material is paper and pulp production, along with a small fraction of cellulose derivative, whereas starches are mainly employed for the production of chemicals including surfactants, binders and additives [4]. Broadly speaking, biomass includes two categories of materials. One is "spontaneous" biomass, based on forest residues (such as dead trees or branches), wood chips, solid waste or any material that can be used for producing heat either by direct burning or after previous conversion into a convenient fuel. The other one includes more valuable materials that can be converted into industrial chemicals or fuels with a high productivity for land surface. Biomass can be grown specifically from numerous types of plants, such as hemp, corn, sugarcane, bamboo, palm and many others. These plants are purposely grown and thus, apart the yield of biomass, the discriminating characteristic is competition with their use as food or of land for food. What is now considered the first generation biomass feedstock was based on edible oil seeds, which, considering the structure, may be a reasonable choice for substituting petrol as precursor of fuels and (some) chemicals, but is no more perceived as practicable from the social point of view since it would subtract land usable for food production. The second generation biomass includes lignocellulose, inedible oil seed crops and waste arising from agricultural and food production. This is a suitable feedstock for the biorefinery, defined by Clark [5] as an *integral unit* that converts biological *nonfood feedstock* via different technologies (including biochemical and thermochemical processes) into a range of useful products including chemicals [6], fuels [7] polymeric materials [8] and energy. However, land requirement is the most important parameters for defying biomass impact. As far as the energetic issue is concerned, the biofuel yield per hectare depends on the chemical produced and ranges from biodiesel, which requires more than four hectares to fuel a single vehicle, to sugarcane ethanol that requires half a hectare [9]. As for the production of chemicals is concerned, the number of initial building blocks must be limited, in analogy with the approach adopted in the case of the petrochemical refinery.

4.2 Chemicals from the Biorefinery

Thus, biomass (in particular lignocellulosic biomass) has been used as a source for oxygenated and non oxygenated bulk chemicals such as hydrocarbons, alcohols, aminoacids (aspartic, glutamic) and acids (including dicarboxylic acids such as fumaric, malic, succinic and 2,5-furan dicarboxylic acids) [5]. Carbohydrates such as mono- or di-saccarides (e.g. glucose, fructose and lactose) can be themselves considered as useful platform chemicals, since they can be easily obtained in high purity from different renewable sources and are successfully employed in sugar chemistry or as precursor to generate most of the other platform molecules indicated by the US department of Energy in 2004 [10] via fermentation methods. In 2010 Bozel and Petersen [11] re-defined the criteria for the selection of biobased platform chemicals from carbohydrates on the basis of recent advances in biotechnology and chemical technology. The new criteria chosen for the selection included, among others, (a) the potentialities of the compound as a direct substitute for petrochemicals and as a primary building block in biorefinery, (b) the use of the compound in a technology that can be applied for the production of several structures and to high volume products, (c) the commercial availability of the product and the attention gained by the compound in literature and obviously (d) the easy production of the candidate from bio renewable feedstock. These criteria allowed for the inclusion of new compounds in the top list, as well as the exclusion of molecules such as glutamic acid (that is considered an end product of chemical industry rather than a useful building block) and glucaric acid (for which an eco-sustainable commercial route starting from renewable feedstock is not yet available). A list of the molecules examined in 2004 and 2010 is summarized in Table 4.1. Among the suitable approaches to convert biomass into valuable products, biochemical methods (both fermentation and biocatalytic) are probably the best choice for green chemistry, in terms of environmental acceptability, selectivity and efficiency. In fact, this approach allows the adoption of mild (physiological) reaction conditions and of an aqueous medium, along with the dual advantage of preventing a large consumption of metal-based catalyst, while at the same time operating through a shorter synthetic sequence.

4.2.1 Alcohols, Glycols and Epoxides

A typical bulk fermentation processes, (bio)ethanol[1] can be produced from a large range of biomass feedstock, including sugar, and starch crops (such as sugar cane and corn), or lignocellulosic feedstock (mainly wood and agricultural waste) and the production steps, namely milling, pressing and cooking the raw materials,

---

[1]The prefix *bio* for biocatalytic products from renewable resources seems to be strongly established in the literature, although this has obviously no chemical meaning.

Table 4.1 Top list of
molecules considered as
platform chemicals according
to Refs. [10, 11]

Compounds
Acids
Succinic acid (since 2004)
Fumaric acid (since 2004)
Malic acid (2004, not included in the 2010 list)
Lactic acid (since 2004)
Glucaric acid (2004, not included in the 2010 list)
Levulinic acid (since 2004)
Itaconic acid (2004, not included in the 2010 list)
3-hydroxypropionic acid (since 2004)
2,5-furan dicarboxylic acid (since 2004)
Aminoacids
Glutamic acid (2004, not included in the 2010 list)
Aspartic acid (2004, not included in the 2010 list)
Aldehydes, ketones
3-hydroxypropionaldehydes (since 2004)
3-hydroxybutyrolactone (2004, not included in the 2010 list)
Furfural (since 2004)
5-hydroxymethylfurfural (since 2004)
Alcohols
Ethanol (since 2004)
Xilitol (since 2004)
Arabinol (2004, not included in the 2010 list)
Sorbitol (since 2010)
Glycerol (since 2004)
Hydrocarbons
Isoprene (since 2004)
Bioderived hydrocarbons (since 2004)

liquefaction (that consists in the conversion of starchy raw materials into a fermentable mash via hydrolysis), saccharification (hydrolysis of the short chains molecules arising from the liquefaction step to give glucose), fermentation (in which ethanol is produced in from 7 to 15 % yield) and distillation, are similar, independently on the nature of feedstock. The production of ethanol from ligno-cellulosic materials as the feedstock is more demanding, both from the economical [12] and from the energetic point of view [13], due to the required conversion of cellulose into glucose, but remains the preferred path, since there is no competition with the use as food. The large scale production of cellulosic ethanol started in 2008, and was calculated to have an E-factor value of 42 kg kg^{-1}, but this was reduced to a mere 1.1 kg kg^{-1} [14] when the main components of waste, that is water and carbon dioxide were not included in the calculation. This choice was motivated by the fact that the process only uses materials from renewable sources and thus no net greenhouse gas is added to the atmosphere. However, it is doubtful that this is appropriate, because the large amount of polluted water can't be reused as such and rather requires an appropriate treatment for reducing the concentration of organic contaminants that are by-produced. The production of ethanol from cellulose has been the subject of several environmental and economic analyses, and LCA has been often used for comparing ethanol production technologies and different materials as feedstock [14]. A recent study by Guo et al. reported the LCA modeling of fuel E100 (100 % bioethanol) and E85 (a 85 % bioethanol/15 gasoline blend) when poplars from different EU countries were used as feedstock. The obtained results showed that the fuels examined were characterized by environmental impact scores (including global warming potential, ozone depletion potential and photochemical oxidation potential) from 10 up to 90 % lower than petrol. In addition, the scenario could be strongly improved (with a further reduction of up to 50 %) by taking into account the use of hybrid poplar clones with higher biomass yields and improved cell wall accessibility [15]. Among higher alcohols, interest around 1-butanol (**1**) is increasing, due to its application in surface coatings or as fuel with higher energy density (it contain 25 % more energy than ethanol) and lower volatility than ethanol [16] 1-Butanol is obtained from petrochemical feedstock through the hydroformylation of propene with carbon monoxide and hydrogen (oxo-synthesis), followed by hydrogenation of the obtained 1-butanal. As for the fermentation pathway, the acetone–butanol–ethanol (ABE) fermentation of glucose containing raw materials such as corn and sugar cane derivatives wheat straw and municipal solid waste (MSW) is one of the largest biotechnological processes performed at present (see Scheme 4.1) [17], beaten in volume only by ethanol fermentation. The sustainability of the biobased process has been analyzed by Uyttebroek et al. that calculated an E-factor of 3.7 kg kg^{-1} (material efficiency: 23 %). However, if by-products arising from the process, namely acetone, ethanol, proteins and fiber are valorized, the E factor drops down to 1.4 kg kg^{-1}, with a material efficiency of 42 % [18]. At present, this performance is not yet competitive with that obtained for the highly atom economic (AE = 1) oxosynthesis, characterized by a E-factor = 0.1 kg kg^{-1}, which is the ideal value for bulk chemicals. Furthermore, the energy supply required by the fermentative process is about twice that required by the petrochemical counterpart [18].

Analogously, 2,3-butylene glycol (2,3-BD) finds application as a solvent, liquid fuel (its heating value is 27,200 J g^{-1}, comparable to that of ethanol and methanol

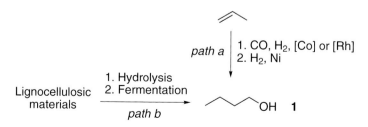

Scheme 4.1 Oxo-synthesis (*path a*) and fermentative (*path b*) route to n-butanol **1**

that are 29,100 and 22,100 J g^{-1} respectively), in monomers and as an additive in polymer chemistry. 2,3-BD can be obtained by fermentation of all of the sugars commonly found in hemicellulose and cellulose hydrolyzates including glucose and cellobiose, with a theoretical maximum yield of glycol from sugar of 0.50 kg kg^{-1} of sugar [19]. Non natural chemical 1,4-butandiol is used as monomer in the production of more than 2.5 billion tons annually of polymers. It can be produced from both nonrenewable and renewable feedstock, the latter involving the fermentation of glucose to succinic acid, followed by catalytic reduction. Recently, Yim et al. reported the production of 1,4-butandiol (in concentration up to 18 g l^{-1}) from renewable carbohydrates (including glucose, xylose, sucrose and biomass derived mixed sugar streams) by means of chemically engineered *Escherichia coli* [20]. LCA analysis demonstrated that the biological pathway has a promising potential for impact in several categories, such as global warming, water use and the disposal of solid waste [21]. Propylene glycol (**2**, 1,2-PO) is another building block with an annual global demand exceeding the 1 million tons. The most important industrial production of racemic 1,2-PDO is based on the hydration of propylene oxide (which in turn arises from propylene, see below, Scheme 4.3), whereas a biobased pathway involving glycerol as starting material is also employed, though to a lesser extent. The main drawback of the petrochemical pathway is the large amount of brine effluent produced (up to 30–60 kg kg^{-1} of employed propylene oxide). On the other hand, glycerol accounts for about 10 wt% of the output of a triglyceride-based biodiesel production process, where it is obtained in 85 % purity. The reaction pathway involves the catalytic dehydration of crude glycerol to the corresponding acetol intermediate. Ensuing hydrogenation (H$_2$, 10 bar, T = 200 °C) affords the desired product in high yields (>70 %). Starting from 2012, BASF and Oleon operated a new manufacturing plant in Belgium for the production of propylene glycol in the 20 kT for year scale. A comparative analysis carried out by Marinas et al. highlighted the significant strength of the bio-based route. This is characterized by a lower E-Factor (0.24, vs. 1.01 kg kg^{-1} calculated for the propylene oxide route) and an energy efficiency value significantly higher (about 70 %) than the corresponding petrochemical process [22] (Scheme 4.2).

Furthermore, despite the lower cost of petrochemical derived glycol, the bio-based route has considerable margins for improvement, since it can employ both technical grade glycerol (97 % purity) and crude grade glycerol (85 %) [22].

4.2.2 Aldehydes, Ketones and Derivatives

Acid or metal catalyzed dehydration of monosaccharides has been exploited for the production of platform chemicals furfural and 5-hydroxymethylfurfural (**3**, HMF) [23]. As an example, dehydratation of pentose (mainly xylose) deriving from hemicellulose is the elective process for the industrial production of furfural

Scheme 4.2 Synthesis of
1,2-propilen glycol **2** via
petrochemical (*path a*) and
biomass derived (*path b*)
feedstock [22]

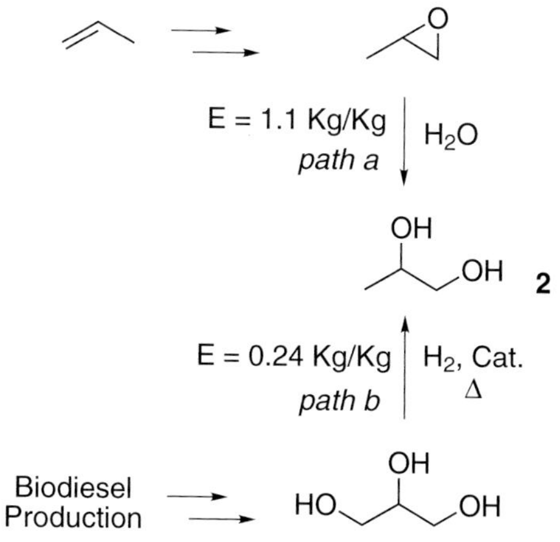

that is obtained industrially in the scale of ca. 200,000 tons year^{-1}, whereas
5-hydroxymethylfurfural can be obtained via dehydration of different hexoses
including fructose [24], glucose and xilose [25]. Both compounds have found broad
application in the synthesis of solvents, chemicals (including furan, tetrahydrofuran,
levulinic acid and γ-valerolactone), fuels [26], fuel additives [27] and polymers.
Furfural (0.1 kg for kg ethanol produced) and **3** are also by-products of production
of lignocellulosic ethanol [23]. A furfural derivative, 5-(chloromethyl)furfural
(**4**, CMF) is obtained in a single step from the treatment of sugars, cellulose, or raw
cellulosic biomass in a biphasic system (aqueous HCl, dichloroethane) in up to
90 % yield [28]. By tuning the reaction conditions, **4** can be in turn converted to
5-hydroxymethylfurfural, levulinic acid (**5**, LA) or the corresponding ethyl levulinate (**6**, ELA). By this approach LA can be thus obtained in 79–87 % overall yield
from carbohydrates (Scheme 4.3).

An efficient multistep synthesis of antiulcer drug ranitidine (ZANTAC) has been
reported by Mascal and Dutta [29]. Thus, coupling of **4** with *N*-acetylcysteamine
afforded the key thioethylamino fragment **7**. Reductive amination of 7 to amine **8**,
hydrolysis of the acetyl group and final coupling with thioderived
1-methylthio-1-methylamino-2-nitroethylene afforded ranitidine **9** in 68 % overall
yield and excellent overall Atom Economy (AE = 0.563). Analogously, the pho-
todynamic drug D-aminolevulinic acid (**11**, ALA) has been obtained starting from
the same biomass derived substrate through a three step process involving conversion into 5-(azidomethyl)furfural **10** and photosensitized oxidation followed by
catalytic hydrogenation. Depending on the raw material used to obtain **4** (including
corn stover, cellulose, and sucrose, see Scheme 4.4) the overall yield of ALA
ranged from 55 to 61 % [30].

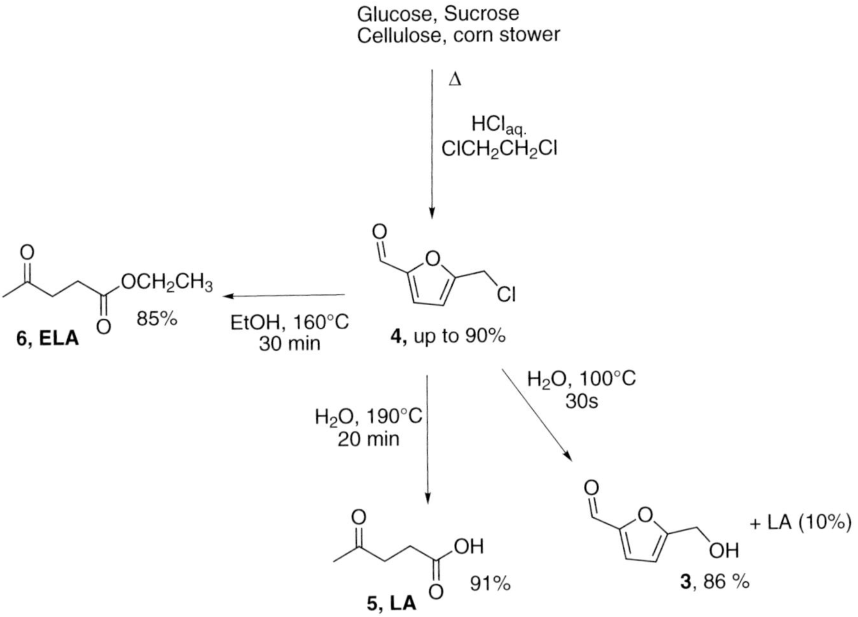

Scheme 4.3 Platform chemicals from 5-(chloromethyl)furfural

4.2.3 Carboxylic Acids and Derivatives

Microbial production of succinic acid, a C_4 building block of strategic importance in the future development of a biorafinery-based chemistry, has been developed by different companies including Myriant Technologies and Reverdia. The latter exploits genetically modified anaerobic bacteria and carries out the fermentation at pH = 3. Thus, succinic acid is directly produced, rather than salts that would have to be neutralized at the end. On the other hand, in the Myriant process, ammonium sulfate arising from the neutralization of the succinate is obtained as by-product. However, water could be partially recovered from waste and furthermore ethanol was produced as a valuable co-product. Pinazo et al. compared the two patented proposal for biosuccinic acid production with the petrochemical route, which involves hydrogenation of maleic anhydride (carried out at high pressure (up to 0.5 MPa) and a relatively high temperature (120–180 °C) in the presence of Ni- or Pd-based catalysts) followed by hydrolysis of the resulting succinic anhydride. The above mentioned valorization of by-products and of coproduced ethanol played a key role in counterbalancing waste production. This resulted in a reduction of the calculated E-factor from 31 down to 6.8 (when ammonium sulfate was valorized, as in the case of Myriant's proposal) and to 1.5–0.7 kg kg^{-1}, if water too was recovered, the final value depended on the feedstock used, to be compared with a calculated E-factor of 0.319 kg kg^{-1} (water was not considered as a waste, since it

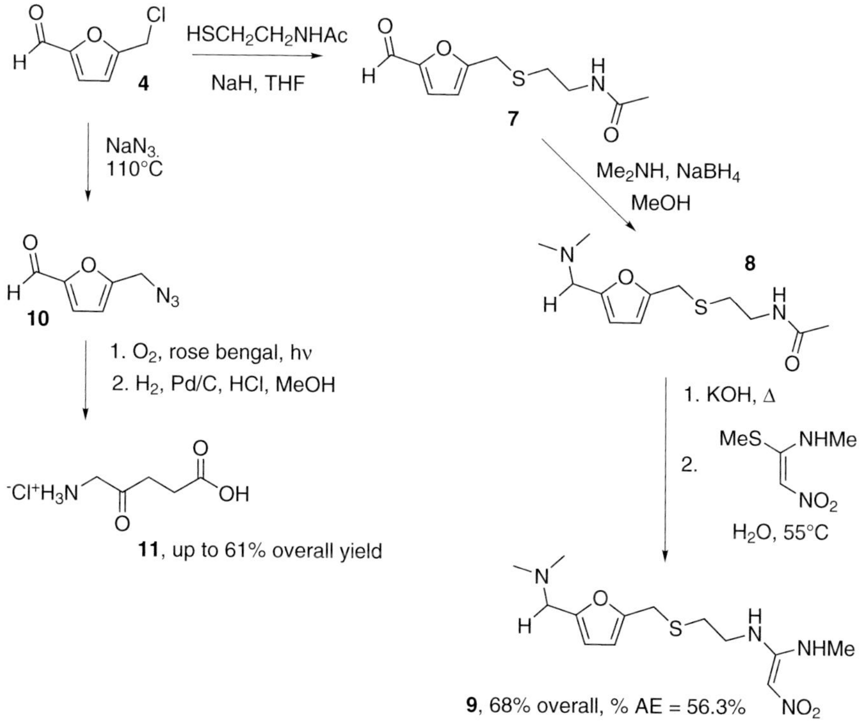

Scheme 4.4 Sinthesis of ranidine (**9**) and D-aminolevulinic acid (**11**) from 5-(chloromethyl)fur-fural (**4**)

was recycled in the process). Furthermore, the energy efficiency of the biocatalytic route to succinic acid was significantly higher than that of its petrochemical counterpart (23 %, with a 356 kWh required energy supply), resulting in a value of ca 30 and 53 % for Myriant and Reverdia's proposal, respectively. The competitiveness of bio-derived succinic acid is evidenced also from the analysis of the production cost, that is ca. 1040 €/MT (in the less favorable conditions), that is lower than the current cost of the petrochemical way, and indeed close to the price of the starting substrate maleic anhydride used in that case [31].

Lactic acid (**12**, LA) and related compounds have found a plethora of applications ranging from food chemistry to the synthesis of biodegradable polymers, such as poly(lactic) acids or polylactide (PLA), produced by NatureWorks (Cargill/Dow LLC) in a 140,000 ton year^{-1} scale. As in the case of succinic acid, the environmental performance of alternative approaches, chemical synthesis and microbial fermentation, have been compared. The feedstock for the production of **12** is a pretreated biomass, rich either in starch or in lignocelluloses. This is hydrolyzed in the presence of acid or of a hydrolytic enzyme mixtures. The resulting mixture is then sent to the fermentative step. After fermentation, filtration and concentration of the crude mixture are carried out and yield a 50 % w/w solution of lactate, that is

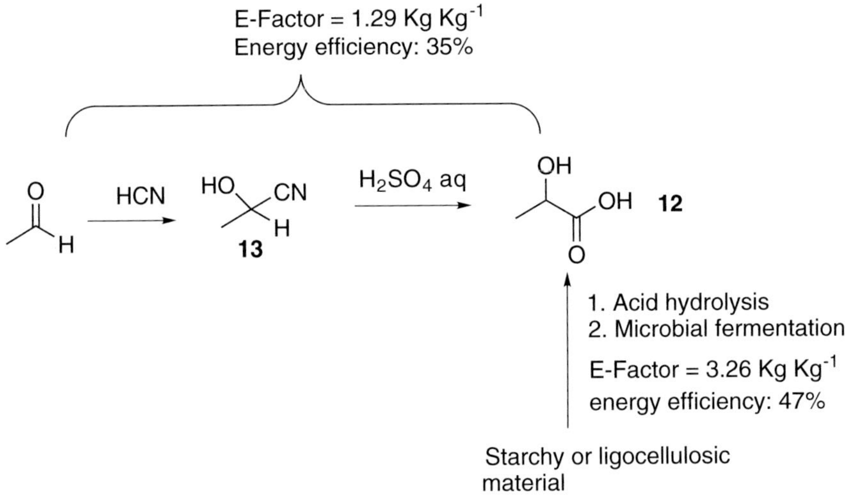

Scheme 4.5 Chemical and biochemical synthesis of lactic acid **12**

treated with sulfuric acid to obtain the free acid (see Scheme 4.5). Conversion of the acid into the corresponding methyl ester and ensuing distillation is the preferred purification step. The alternative chemical synthesis is the reaction of acetaldehyde with hydrogen cyanide followed by acid catalyzed hydrolysis of the obtained lactonitrile (**13**, T = 100 °C). Furthermore, LA is chemically synthesized as a D/L racemic mixture, whereas bio-lactic acid can be selectively produced either as levo- or as dextro-rotatory enantiomer. The green metrics assessment carried out by Juodeikiene et al. showed that the production of LA from biomass resulted in an energy efficiency up to 47 % and likewise to a 17 % lower overall production cost (1.0 vs. 1.2 k\$ ton^{-1}), despite the less satisfactory mass efficiency (48 % vs. 75 % for the chemical process) and the higher value of E-factor (3.26 vs. 1.29 kg kg^{-1}) [32].

As hinted above, ester ethyl lactate is one of the most promising ecosustainable compounds and finds a large variety of applications, such as food additive and solvent in organic synthesis and in polymer chemistry, with a worldwide market of about 30 million pounds per year. It is produced industrially in a racemic mixture via acid catalyzed esterification of ethanol with lactic acid, which is in turn obtained from biomass fermentation. Mueller reported that a significant reduction of greenhouse emissions (up to 37 %) can be obtained when a corn ethanol plants is integrated with an ethyl lactate coproduction plant, based on a life cycle analysis approach [33]. An E-factor of 1.4 kg kg^{-1} has been reported by Petrides et al. for another bulk fermentation product, citric acid. In this case, 75 % of the waste is accounted for by calcium sulfate, arising from the neutralization of calcium hydroxide added to the batch fermentation to control the pH. The low environmental contribution of the salt brings the E-factor well within the range of <1–5 kg kg^{-1}, typical of bulk petrochemicals. However, inclusion of water in the calculation has again a large effect and an E factor of 17 kg kg^{-1} was calculated under these conditions [34].

4.2.4 Hydrocarbons

As for non-oxygenated products, the production of light alkenes such as ethylene, propylene and isoprene via microbial paths has been investigated in depth in recent years [35]. Endothermic dehydration of bioethanol on alumina or silica-alumina based catalysts is currently the most efficient route to bioethylene (99 % conversion yield, 97 % selectivity). LCA analyses pointed out the environmental advantages of bioethylene production with respect to its petrochemical counterpart, with an energy saving of 19 GJ per ton of product and a 40 % reduction in the emitted CO_2. Isoprene is a key chemical in different industrial productions, the most important of which is the production of synthetic rubber (cis-polyisoprene) in tire manufacturing. Isoprene is the main component of the C_5 cracked fraction deriving from the pyrolysis of hydrocarbons to ethylene. On the other hand, isoprene is also biologically produced by different bacterial strains, both Gram-positive and Gram-negative, and the development of modified bacteria for efficient isoprene production is being considered by different tire companies including the Goodyear Tire and Rubber Company and Michelin. The assessment of the different routes to this key product has been recently reported [35]. When using modified *Escherichia coli* bacteria and glucose as the carbon source, comparable results in terms of E-factor (see Table 4.2), material efficiency and cost were found, with the latter parameters slightly favorable to the petrochemical alternative. Furthermore, the use of waste rich in carbohydrates is essential to avoid land use for aims different from food production.

When small-volume fermentations are considered, E-factors tend to be very high. This is the case of pharmaceutical bioprocesses, as exemplified by the E-factor of about 6600 kg kg^{-1} (E = 50,000 kg kg^{-1} if water is included) observed for the production of recombinant human insulin by non pathogenic biochemically engineered *Escherichia coli*. In this case, the main contributors to the E-factor value were urea, acetic, formic and phosphoric acid, guanidine hydrochloride, glucose, acetonitrile, along with inorganic salts such as sodium chloride and hydroxide [36].

Table 4.2 Environmental parameters for the production of isoprene

Parameters	Petrochemical derived	Biomass derived
E-factor (kg kg^{-1})	1.50	1.19
Material efficiency	0.40	0.46
Energy efficiency	0.29	0.55
Production cost (€ ton^{-1})	4333	4950
Land use (ha ton^{-1})	0	0.8

References

1. Metzger JO (2004) Agenda 21 as a guide for green chemistry research and a sustainable future. Green Chem 6:G15–G16
2. Balzani V, Credi A, Venturi M (2008) Photochemical conversion of solar energy. ChemSusChem 1:26–58
3. Field CB, Behrenfeld MJ, Randerson JT, Falkowski P (1998) Primary production of the biosphere: integrating terrestrial and oceanic components. Science 281:237–240
4. Röper H (2002) Renewable raw materials in Europe—industrial utilization of starch and sugar. Starch/Stärke 54:89–99
5. Clark JH, Luque R, Matharu AS (2012) Green chemistry, biofuels, and biorefinery. Annu Rev Chem Biomol Eng 3:183–207
6. Sheldon RA (2014) Green and sustainable manufacture of chemicals from biomass: state of the art, Green Chem 16:950–963; Gallezot P (2007) Process options for converting renewable feedstocks to bioproducts. Green Chem 9:295–302
7. Nigam PS, Singh A (2011) Production of liquid biofuels from renewable resources. Progr Energy Comb Sci 37:52–68
8. Harmsen PFH, Hackmann MM, Bosenergy HL (2014) Green building blocks for bio-based plastics. Biofuels Bioprod Biorefin 8:306–324
9. Sagar AD, Kartha S (2007) Bioenergy and sustainable development? Ann Rev Environ Resour 32:131–167
10. Werpy T, Petersen G (eds) (2004) Top value added chemicals from biomass volume I—results of screening for potential candidates from sugars and synthesis gas, DOE/GO-102004-1992, 1 Aug 2004
11. Bozell JJ, Petersenm GR (2010) Technology development for the production of biobased products from biorefinery carbohydrates—the US Department of Energy's "Top 10" revisited. Green Chem 12:539–554
12. Gnansounou E, Dauriat A (2010) Techno-economic analysis of lignocellulosic ethanol: a review. Bioresour Technol 101:4980–4991
13. Rinaldi R, Schüth F (2009) Design of solid catalysts for the conversion of biomass. Energy Environ Sci 2:610–626
14. see for review: Gnansounou E, Dauriat A (2010) Techno-economic analysis of lignocellulosic ethanol: a review. Biores Technol 101:4980–4991; Singh A, Pant D, Korres NE, Nizami A-S, Prasad S, Murphy JD (2010) Key issues in life cycle assessment of ethanol production from lignocellulosic biomass: challenges and perspectives. Bioresour Technol 101:5003–5012
15. Guo M, Littlewood J, Joyce J, Murphy R (2014) The environmental profile of bioethanol produced from current and potential future poplar feedstocks in the EU. Green Chem 16:4680–4695
16. Jin C, Yao M, Liu H, Leed CF, Ji J (2011) Progress in the production and application of n-butanol as a biofuel. Renew Sustain Energy Rev 15:4080–4106
17. Green EM (2011) Fermentative production of butanol—the industrial perspective. Curr Opin Biotechnol 22:337–343
18. (a) Menon V, Rao M (2012) Trends in bioconversion of lignocellulose: biofuels, platform chemicals and biorefinery concept. Progr Energy Comb Sci 38:522–550. (b) Material efficiency is defined as 1/(E + 1) where one is the sum of all the useful products, see Sheldon RA, Sanders JPM (2015) Toward concise metrics for the production of chemicals fromrenewable biomass. Cat Today 239:3–6
19. Uyttebroek M, Van Hecke W, Vanbroekhoven K (2013) Sustainability metrics of 1-butanol. Cat Today. doi:10.1016/j.cattod.2013.10.094
20. Yim H, Haselbeck R, Niu W, Pujol-Baxley C, Burgard A, Boldt J, Khandurina J, Trawick JD, Osterhout RE, Stephen R, Estadilla J, Teisan S, Schreyer HB, Andrae S, Yang TH, Lee SY, Burk MJ, Van Dien S (2011) Metabolic engineering of *Escherichia coli* for direct production of 1,4-butanediol. Nat Chem Biol 7:445–452

21. Curran MA (2000) Life cycle assessment: an international experience. Environ Prog 19:65–71; US Environmental Protection Agency (1997) Streamlined life-cycle assessment of 1,4-butanediol produced from petroleum feedstocks versus bio derived feedstocks. National Risk Management Research Laboratory, Cincinnati, Ohio
22. Marinas A, Bruijnicx P, Ftouni J, Urbano FJ, Pinel C (2015) Sustainability metrics for a fossil- and renewable-based route for 1,2-propanediol production: a comparison. Cat Today 239:31–37
23. Tong X, Ma Y, Li Y (2010) Biomass into chemicals: conversion of sugars to furan derivatives by catalytic processes. Appl Cat A Gen 385:1–13
24. See for instance Moreau C, Durand R, Razigade S, Duhamet J, Faugeras P, Rivalier P, Ros P, Avignon G (1996) Dehydration of fructose to 5-hydroxymethylfurfural over H-mordenites. Appl Catal A 145:211–224. See for other procedures for the synthesis of 5-hydroxymethylfurfural see: Zhao H, Holladay JE, Brown H, Zhang ZC (2007) Metal chlorides in ionic liquid solvents convert sugars to 5-Hydroxymethylfurfural. Science 316:1597–1600; Hu S, Zhang Z, Song J, Zhou Y, Han B (2009) Efficient conversion of glucose into 5-hydroxymethylfurfural catalyzed by common Lewis acid $SnCl_4$ in an ionic liquid. Green Chem 11:1746–1749; Su Y, Brown HM, Huang X, Zhou X-d, Amonette JE, Zhang ZC (2009) Single-step conversion of cellulose to 5-hydroxymethylfurfural (HMF), a versatile platform chemical. Appl Cat A Gen 361:17–122; Rosatella AA, Simeonov SP, Frade RFM, Afonso CAM (2011) 5-Hydroxymethylfurfural (HMF) as a building block platform: biological properties, synthesis and synthetic applications. Green Chem 13:754–793
25. Chheda JN, Romàn-Leshkov Y, Dumesic JA (2007) Production of 5-hydroxymethylfurfural and furfural by dehydration of biomass-derived mono- and poly-saccharides. Green Chem 9:342–350
26. Lange JP, van der Heide E, van Buijtenen J, Price R (2012) Furfural, a promising platform for lignocellulosic biofuels. ChemSusChem 5:150–166; Huber GW, Cheda JN, Barrett CJ, Dumesic JA (2005) Production of liquid alkanes by aqueous-phase processing of biomass-derived carbohydrates. Science 308:1446–1450
27. Schmidt LD, Dauenhauer PJ (2007) Chemical engineering: hybrid routes to biofuels. Nature 447:914–915
28. Mascal M, Nikitin EB (2010) High-yield conversion of plant biomass into the key value-added feedstocks 5-(hydroxymethyl)furfural, levulinic acid, and levulinic esters via 5-(chloromethyl)furfural. Green Chem 12:370–373
29. Mascal M, Dutta S (2011) Synthesis of ranitidine (zantac) from cellulose-derived 5-(chloromethyl) furfural. Green Chem 13:3101–3102
30. Mascal M, Dutta S (2011) Synthesis of the natural herbicide d-aminolevulinic acid from cellulose-derived 5-(chloromethyl)furfural. Green Chem 13:40–41
31. Pinazo JM, Domine ME, Parvulescu V, Petruca F (2015) Sustainability metrics for succinic acid production: a comparison between biomass-based and petrochemical routes. Cat Today 239:17–24
32. Juodeikienea G, Vidmantienea D, Basinskienea L, Cernauskasa D, Bartkieneb E, Cizeikienea D (2015) Green metrics for sustainability of biobased lactic acid from starchy biomass vs chemical synthesis. Cat Today 239:11–16. For other synthesis of lactic acid from biomass see for instance: Ramírez-López CA, Ochoa-Gómez JR, Gil-Río S, Gómez-Jiménez-Aberasturi O, Torrecilla-Soria J (2011) Chemicals from biomass: synthesis of lactic acid by alkaline hydrothermal conversion of sorbitol. J Chem Technol Biotechnol 86:867–874; Wang Y, Deng W, Wang B, Zhang Q, Wan X, Tang Z, Wang Y, Zhu C, Cao Z, Wang G, Wan H (2013) Chemical synthesis of lactic acid from cellulose catalysed by lead(II) ions in water. Nat Commun 4. doi:10.1038/ncomms3141
33. Mueller S (2010) Life cycle analysis of ethyl lactate production and controlled flow cavitation at corn ethanol plants. PhD thesis, University of Illinois at Chicago
34. see for review Ladygina N, Dedyukhina EG, Vainshtein MB (2006) A review on microbial synthesis of hydrocarbons. Process Biochem 41:1001–1014

35. Matos CT, Gouveia L, Morais ARC, Reis A, Bogel-Łukasik R (2013) Green metrics evaluation of isoprene production by microalgae and bacteria. Green Chem 15:2854–2864; Morais ARC, Dworakowska S, Reis A, Gouveia L, Matos CT, Bogdał D, Bogel-Łukasik R Chemical and biological-based isoprene production: green metrics (2015) Cat Today 239:38–43
36. Sheldon R (2011) Reaction efficiencies and green chemistry metrics of biotransformations. In: Tao J, Kazlaukas R (eds) Biocatalysis for green chemistry and chemical process development. Wiley, Hoboken

Chapter 5
The Solvent Issue

Abstract Many issues encountered in green chemistry have to do with the solvent. This is usually by far the most abundant component of the mixture and determines the course of the reaction through its physical characteristics as well as the separation and the purification of the end product, the recovery of the catalyst and its own recovery and reuse or disposal. The best choice is discussed.

Keywords Green solvents · Solvent guide · Recovery · Fluorous systems · Supercritical fluids · Ionic liquids · Solventless process

Many issues encountered in green chemistry have to do with the solvent. This is usually the main component of the reaction mixture and its physical characteristics for a large part determine the course of the reaction, as well as the mode of separation and purification of the end products, the recovery of the catalyst and its own recovery and reuse or disposal. Furthermore, solvents play a key role in other steps of the production stream including reactant and product transport during the various stages of its life. An improvement can be looked for by adopting solvents that perform better towards environment, health, safety, by introducing "bio" solvents (that is prepared from renewable resources, not from fossil feedstock), as well as by substituting organic solvents by supercritical fluid not harmless to the environment, or by using non volatile solvents such as ionic liquids.

Solvents of petrochemical origin are classified on the basis of the assessment of the full life-cycle, plus either recovery or disposal. The order of preference is about the same for both choices, because the credits arising from recovery, which give the main contribution in the first option, are determined by the environmental impact of the production of that solvent from oil, which is roughly proportional to the energy recovery from incineration, which predominates in the other option. A combined LCA/EHS analysis of the environmental impact resulting from the production of 26 common solvents and their use in synthesis has been carried out by Capello et al. [1]. In particular, assessment of the solvent with the Environmental Health Safety tool provides a total score for the overall hazards related with the employment of solvents (including persistency, water and air hazards, acute and chronic toxicity), and also points out to specific aspects that are critical for each medium and the

© The Author(s) 2016
A. Albini and S. Protti, *Paradigms in Green Chemistry and Technology*,
SpringerBriefs in Green Chemistry for Sustainability,
DOI 10.1007/978-3-319-25895-9_5

precautionary technical measures that have to be adopted when using it in a chemical process. From this point of view, methanol and ethanol (produced with a low environmental impact) on one hand and hexane and heptane (that have a high net calorific value and thus high environmental credits for energy recovery) on the other are the preferred choice, while isopropanol, ethyl acetate or acetonitrile are discouraged, because of the relatively high impact of the production, not compensated by the low calorific value. As predictable, methanol- or ethanol-water mixtures are preferred to pure alcohols. See Fig. 5.1.

The GSK group has carried out an in depth exploration of this issue and has made available a detailed guide to the use of solvents. In the 2011 version, 110 solvents have been compared with regard to disposal, environmental and health impact, as well as flammability and explosion risk, reactivity and stability, life cycle (obviously including the impact incurred in the preparation of the solvent), current or potential regulations, possibility and cost of recovery (as well as associated risk). High (>120 °C) or low (<40 °C) boiling point derivatives have been discarded. It is clearly advantageous to arrive at the best decision as early as possible during the development of a synthetic method. Thus, it is important that at the explorative level, medicinal chemists or analytic chemists have the appropriate information easily available. As noted by the GSK group, this promotes a change of mind-set among chemists involved in the pilot project and help in exerting their creativity in this field. In order to facilitate the choice, a simplified table has been prepared, where solvents are characterized as presenting only a few, some or major issues [2]. As for purification (and recovering) of solvents, it should be taken in account that *nanofiltration*, which gives yields comparable to those obtained from distillation but requires much less energy, at least in the initial runs, is reasonably expected to play a much more important role in solvent recovery in the near future (see Fig. 5.2).

Another point is the preparation of *solvents from bioresources* rather than petrol chemistry, although bio-derived solvents obviously have the same limitations that their petrochemical counterpart [3]. On the other hand, not all of the traditional organic solvents can be obtained via the available bio-derived feedstock. 2-Methyl tetrahydrofuran can be produced from levulinic acid and 2-furaldehyde and has been exploited as a valid substitute of THF and dichloromethane. Monobasic ester ethyl lactate is a high boiling polar protic solvent industrially produced as a racemic mixture from ethanol and lactic acid, with water as the only by-product. Ethyl lactate found application as wood polystyrene and metals coating as well as paint stripper, replacing solvents such as *N*-methyl pyrrolidone (NMP), acetone, toluene and xylene [4]. The use of other esters such as isoamyl derivatives from fused oil have been also considered [4b], along with glycerol derived solvents [5].

Water can be employed in both liquid [6] and supercritical state. The main industrial application for the liquid state is the Ruhrchemie/Rhone-Poulenc "oxo" process for the hydroformylation of propylene to n-butanal, where a water soluble rhodium(I) complex of trisulfonated triphenylphosphine (tppts) as the catalyst and an aqueous/organic biphase system as the medium are used (compare Sect. 3.3) [7].

(a)

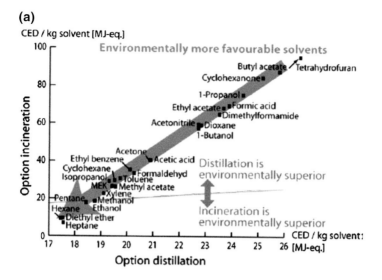

(b)

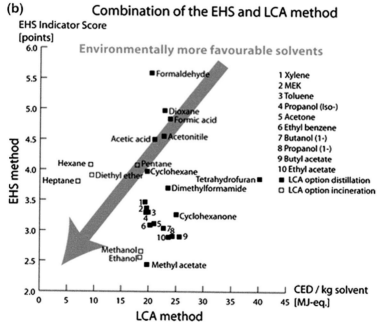

Fig. 5.1 a LCA of the treatment options, incineration and distillation, for 26 common solvents. The results were calculated using the Ecosolvent-Tool. **b** Environmental assessment of 26 organic solvents: combination of the EHS method with the LCA method. Reprinted with permission from Ref. [1]

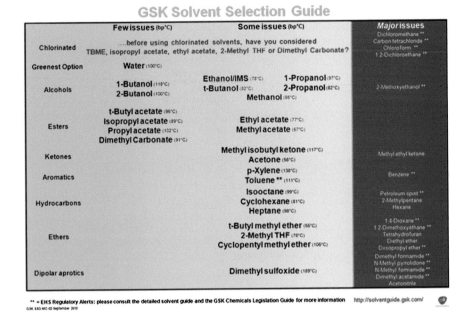

Fig. 5.2 A quick reference solvent guide for chemical scientists according to GSK guidelines. Reprinted with permissions from Ref. [2]

On the other hand, neat water was found to be a promising solvent for heterogeneous catalyzed processes. As an example, in 2013 Kumar et al. reported the atom economic synthesis of substituted phenylacetimidamides **1** via room temperature ZnO nanoparticles (NPs) catalyzed Ugi reaction (NPCR) in aqueous medium [8] (an example in Scheme 5.1). Hydrolysis of the iminoether moiety with I_2–SDS–water system then yielded the corresponding 2-amino-2-phenylacetamide **2**.

Most of organic substrates and catalysts are water-insoluble, but the use of the surfactants that form micelles that act as nanoreactors allows to circumvent the solubility issues, thus enabling a wide range of organic synthesis (including cross coupling processes, oxidation and hydroformylation) to be run in water with environmental benefits. Lipshutz et al. recently reviewed the use of engineered surfactant in water for synthesis calculating a decrease in the E-factor value of more than twenty times for Sonogashira, Heck and Suzuki-Miyauara reactions (from 50 down to ca 2 kg kg^{-1}) [6b].

Supercritical water was used as solvent in a plethora of chemical processes namely partial oxidation, rearrangement and elimination [9]. Dehydration of glycerol to several products including propionic acid, acrolein and acetaldehyde on TiO_2 and WO_3/TiO_2 in supercritical water at 400 °C and 33 MPa was carried out in a fixed bed flow reactor (Scheme 5.2) [10]. The use of superheated water in synthesis has been also considered [11].

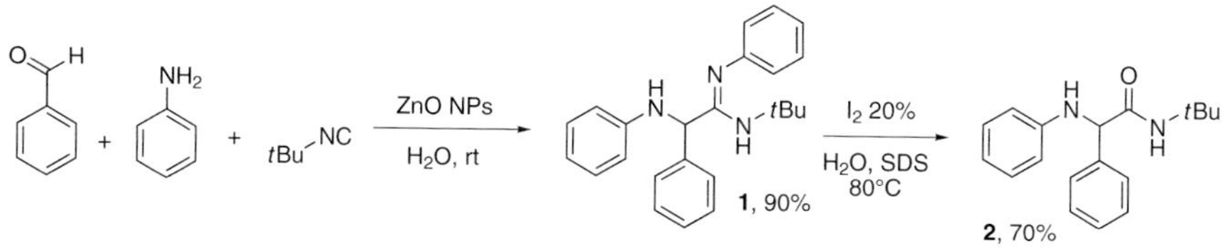

Scheme 5.1 ZnO NPs catalyzed synthesis of phenylacetimidamides via Ugi reaction in water

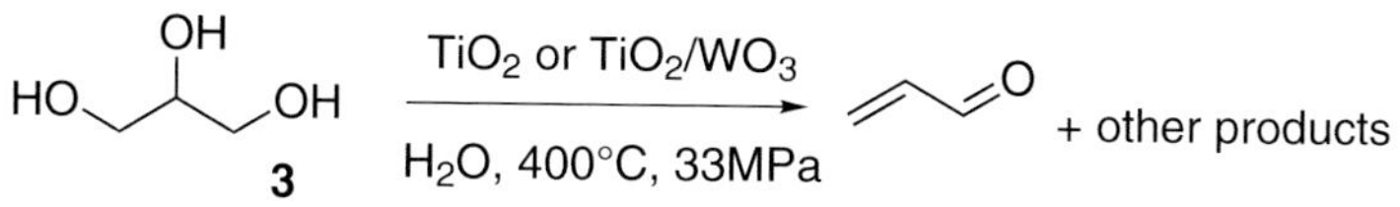

Scheme 5.2 Dehydration of glycerol as source of fine chemicals

Carbon dioxide can be intercepted from industrial waste streams and converted into a supercritical fluid at a temperature and a pressure above 31.1 °C and 73.8 bar respectively [12]. Obviously this is a greenhouse gas, but its use involves no net addition to the atmosphere, as it is "borrowed" and then given back. The gas-like diffusivity and liquid-like density of scCO$_2$ make it one of the best solvents for extraction processes [13] and polymer chemistry including polymerization and formation of polymer composites and polymer blends [14]. As far as synthesis is concerned, scCO$_2$ and other supercritical fluids have found several applications in heterogeneous catalysis [15].

The negligible vapor pressure exhibited by *ionic liquids* suggests that volatile organic compounds (VOC) emissions arising from traditional organic solvents can be minimized through the use of ionic liquid solvents during synthetic processes in industry. Furthermore, the selection of cations and anions available is virtually endless and gives access to a huge variety of physical properties and environmental characteristics, in order to precisely optimize any single application [16].

An explicative example is the conversion of glucose to 5-(hydroxymethyl)fur-fural (HMF) that has been performed with ytterbium chloride or triflate as catalyst and in alkylimidazolium chlorides as solvent. This is a chromium free alternative for the synthesis of these platform chemicals (Scheme 5.3) [17a].

Ionic liquids have found applications as solvents in industry [16] since December 1996, Eastman Chemical Company had been running an industrial scale process for the isomerisation of 3,4-epoxybut-1-ene (**4**) to 2,5-dihydrofuran (**5**) in a Lewis basic ionic liquid (crotonaldehyde was found in 1 % yield as side-product) [17b] (Scheme 5.4).

The use of *fluorous* biphasic systems have been shown to provide a means of recycling a catalyst-containing reaction medium from which products can be easily extracted. Most common fluorinated solvents include perfluoroalkanes, such as perfluoroheptane, and perfluoroethers. The former are typically immiscible with organic solvents at room temperature. Catalyzed reactions can be carried out by

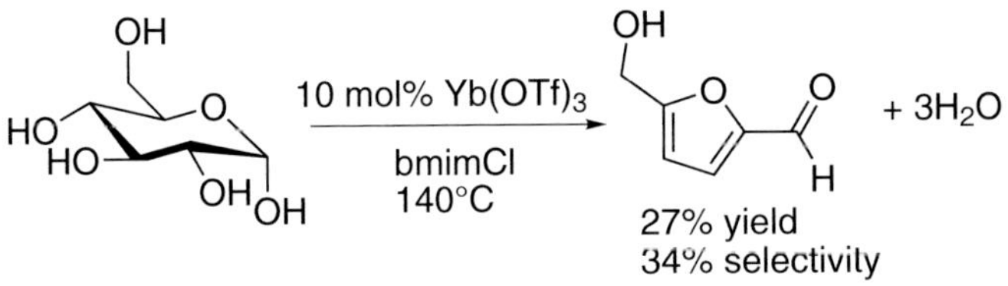

Scheme 5.3 Conversion of glucose to 5-(hydroxymethyl)furfural (HMF) in butyl methyl imidazolium chloride (bmimCl) [17a]

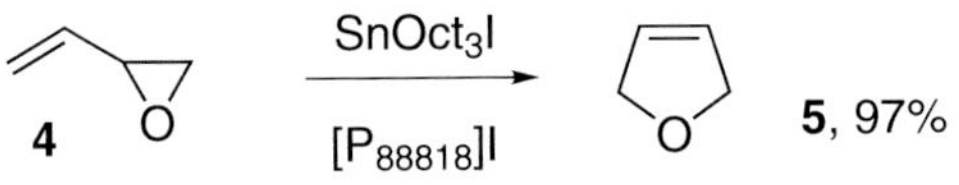

Scheme 5.4 Isomerization of of 3,4-epoxybut-1-ene (**4**) to 2,5-dihydrofuran (**5**)

using a modified catalyst with fluorinated regions in order to improve solubility in these systems and form an easily recyclable catalytic system [18], especially in the case of aerobic oxidations, due the high solubility of oxygen in fluorocarbons. The chemical and biological inertness of perfluorinated solvents makes them appealing, but commercially available compounds of this type are rather expensive if compared with common organic solvents, and not largely applied on industrial scale (at any rate, they are the second largest segment among halogenated solvents) [18, 19] (Scheme 5.5a). As an example, meso-perfluoropropylporphyrin was used as recyclable and stable sensitizer in singlet oxygenation of allylic substrate **6** in preparative scale [19b]. Analogously, oxidation of aliphatic and aromatic aldehydes was carried out by using oxygen (1 bar) in the presence of fluorophilic soluble Ni $(C_7F_{15}COCHCOC_7F_{15})_2$ complex as catalyst [19c] (In Scheme 5.5b the conversion of aromatic aldehyde **8** to acid **9**).

It remains true, however, that the best solvent is no solvent and the use of *solvent-less* procedures is rapidly increasing, particularly in condensation and redox reactions [20, 21]. As an example, 2H-indazolo[1,2-b]phthalazine-triones **10** have been synthesized through a four-component condensation reaction from phthalic anhydride, hydrazinium hydroxide, dimedone or 1,3-cyclohexanedione, and arylaldehydes using PEG-OSO$_3$H [22] as catalyst (Scheme 5.6).

A protocol for the selective fluorination in position 2 of organic 1,3-dicarbonyl substrates under solvent free conditions was proposed by Stavber et al. [23], by using Accufluor^{TM} NFSi (*N*-fluorobenzenesulfonimide) (NFSi in scheme 5.7) as fluorinating agent, the reaction proceeding in the molten eutectic phase of the reactants (Scheme 5.7). Similar results have been obtained in water, by using Selectfluor^{TM} F-TEDA-BF$_4$ (1-chloromethyl-4-fluoro-1,4-diazonia-bicyclo[2.2.2] octane bis-tetrafluoroborate) as fluorine source.

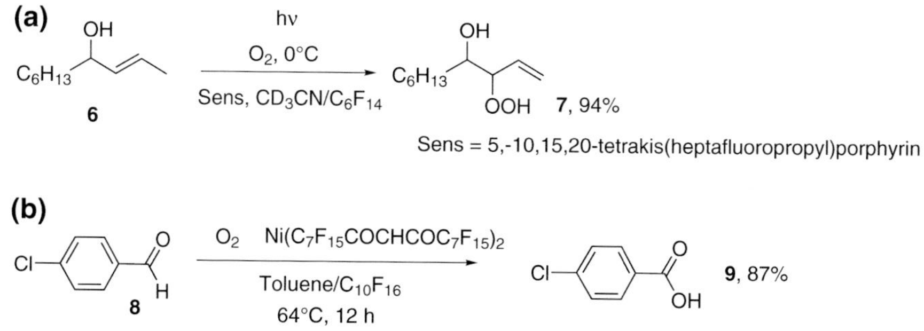

(a)

Sens = 5,-10,15,20-tetrakis(heptafluoropropyl)porphyrin

(b)

Scheme 5.5 Organic reaction in fluorous biphase system [19b,c]

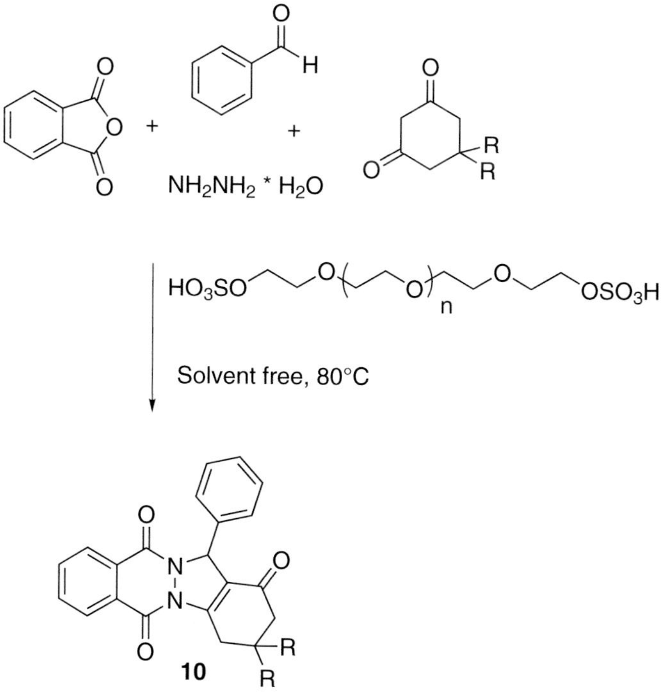

Scheme 5.6 Multicomponent synthesis of **10** under solvent free conditions [22]

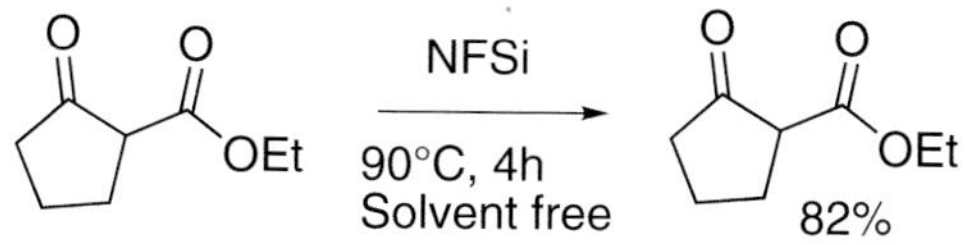

Scheme 5.7 Fluorination of 3-dicarbonyl compounds under solvent free conditions [23]

References

1. (a) Capello C, Fischer U, Hungerbühler K (2007) What is a green solvent? A comprehensive framework for the environmental assessment of solvents. Green Chem 9:927–934. (b) Sugiyama H, Fischer U, Hungerbühler K (2006) The EHS tool. ETH Zurich, Safety & Environmentak Technology group, Zurich 2006. http://www.sust-chem.ethz.ch/tools/EHS)
2. (a) Henderson RK, Jiménez-Gonzalez C, Constable DJC, Alston SR, Inglis GGA, Fisher G, Sherwood J, Binks, SP, Curzons AD (2011) Expanding GSK'solvent selection guide— embedding sustainability into solvent selection starting at medicinal chemistry. Green Chem 13:854–862. (b) An analogous guideline has been compiled for reagents, see Fig 1 in Ch 3
3. Horvàth I (2008) Solvents from nature. Green Chem 10:124–128
4. (a) Pereira CSM, Silva VMTM, Rodrigues AE (2011) Ethyl lactate as a solvent: properties, applications and production processes—a review. Green Chem 13:2658–2671; (b) Bandres M, de Caro P, Thiebaud-Roux, Borredon ME (2011) Green syntheses of biobased solvents. Compt Rend 14:636–646
5. García JI, García-Marína H, Mayoral JA, Pérez P (2010) Green solvents from glycerol. Synthesis and physico-chemical properties of alkyl glycerol ethers. Green Chem 12:426–434
6. (a) Li C-J, Chen L (2006) Organic chemistry in water. Chem Soc Rev 35:68–82; (b) Lipshutz BH, Ghorai S (2014) Transitioning organic synthesis from organic solvents to water. What's your E-Factor? Green Chem 16:3660–3679
7. Cornils S, Wiebus E (1996) Hydration and dehydration of olefins is likewise carried out under these conditions. Recl Trav Chim Pays-Bas 115:211–215
8. Kumar A, Saxena D, Gupta MK (2013) Nanoparticle catalyzed reaction (NPCR): ZnO-NP catalyzed Ugi-reaction in aqueous medium. Green Chem 15:2699–2703
9. Savage PE (1999) Organic chemical reactions in supercritical water. Chem Rev 99:603–621
10. Akizuki M, Oshima Y (2013) Kinetics of glycerol dehydration with WO_3/TiO_2 in supercritical water. Ind Engin Chem Res 51:12253–12258
11. Messina F, Rosati O (2013) Superheated water as solvent in microwave assisted organic synthesis of compounds of valuable pharmaceutical interest. Curr Org Chem 17:1158–1178
12. Gang Z, Huan-Feng J, Ming-Cai C (2003) Chemical reactions in supercritical carbon dioxide. ARKIVOC ii:191–198. Beckman EJ (2004) Supercritical and near-critical CO_2 in green chemical synthesis and processing. J Supercritical Fluids 28:121–191
13. King MB, Bott TR (eds) (2012) Extraction of natural products using near-critical solvents. Springer, Dordrecht
14. Nalawade SP, Picchioni F, Janssen LPBM (2006) Supercritical carbon dioxide as a green solvent for processing polymer melts: processing aspects and applications. Prog Polym Sci 31:19–43
15. Baiker A (1999) Supercritical fluids in heterogeneous catalysis. Chem Rev 99:453–473
16. See for reviews: Welton T (1999) Room-temperature ionic liquids. Solvents for synthesis and catalysis. Chem Rev 99:2071–2083. Hallett JP, Welton T (2011) Room-temperature ionic liquids: solvents for synthesis and catalysis. 2. Chem Rev 111:3508–3576; Zhao H, Malhotra SV (2002) Applications of ionic liquids in organic synthesis. Aldrichimica Acta 35:75–83
17. (a) Stahlberg T, Grau Sørensen M, Riisager A (2010) Direct conversion of glucose to 5-(hydroxymethyl)furfural in ionic liquids with lanthanide catalysts. Green Chem 12:321–325; (b) Plechkova NV, Seddon KR (2008) Applications of ionic liquids in the chemical industry. Chem Soc Rev 37:123–150
18. Sheldon RA (2005) Green solvents for sustainable organic synthesis: state of the art. Green Chem 7:267–278
19. Horvàth IT (1998) Fluorous biphase chemistry. Acc Chem Res 31:641–650. See also (a) Curran DP, Hadida S (1996) Tris(2-(perfluorohexyl)ethyl)tin hydride. A new fluorous reagent for use in traditional organic synthesis and liquid phase combinatorial synthesis. J Am Chem Soc 118:2531–2532. (b) Di Magno SG, Dussault PH, Schultz JA (1996) J Am Chem

Soc 118:5312–5313. (c) Klement I, Lutjens H, Knochel P (1997) Transition metal catalyzed oxidations in perfluorinated solvents. Angew Chem Int Ed Engl 36:1454–1456

20. Hernandez JG, Avila-Ortiz, CG, Juaristi E (2014) Useful chemical activation alternatives in solvent-free organic reactions. In: Knochel P, Molander GA (eds) Comprehensive organic synthesis, 2nd edn, vol 9, pp 287–314. Shah JS, Patel PV, Maheshwari DG (2015) Current perspective on solvent free analysis in pharmaceutical industry. World J Pharm Pharmaceut Sci 4:1770–1780; Cravotto G, Gaudino EC (2014) Oxidation and reduction by solid oxidants and reducing agents using ball-milling. In: Stolle A, Ranu B (eds) Ball milling towards green synthesis: applications, projects, challenges. Roy Soc Chem 2014:58–80

21. Banon-Caballero A, Guillena G, Najera C (2014) Solvent-free enantioselective organocatalyzed aldol reactions. Mini-Rev Org Chem 11:118–128

22. Hasaninejed A, Kazerooni MR, Zareb A (2012) Solvent-free, one-pot, four-component synthesis of 2H-indazolo[2,1-b]phthalazine-triones using sulfuric acid-modified PEG-6000 as a green recyclable and biodegradable polymeric catalyst. Catal Today 196:148–155; see also Vekariya RH, Patel HD (2015) Sulfonated polyethylene glycol (PEG-OSO3H) as a polymer supported biodegradable and recyclable catalyst in green organic synthesis: recent advances. RSC Adv 5:49006–49030

23. Stavber G, Stavber S (2010) Towards greener fluorine organic chemistry: direct electrophilic fluorination of carbonyl compounds in water and under solvent-free reaction conditions. Adv Synth Catal 352:2838–2846

Chapter 6
Process Intensification in Organic Synthesis

Abstract Optimization at the scaling up stage and the engineering of the final process at the stage of commercial process contribute to the environmental role of the process as least as much as the merely chemical issues. 12 Principles of green chemical engineering have been formulated, in part parallel to green chemistry. Process intensification involves not only the better use of the space available in the plant, but also revising previous chemistry to introduce novel reactions simultaneously with the development of new (most often multifunctional) apparatuses, with the only predetermined parameter of the better yield. Intrinsically safer procedures are found that are also economically profitable.

Keywords Process intensification · Scale up · Chemical engineering · Flow systems

6.1   Green Chemical Engineering and Green Chemistry

Serious environmental effects may be due to trace contaminants [1]. When a chemical is selected for industrial production, the scaling up issue must be considered with attention, in order to avoid that the product itself and any other component of the mixture can cause a significant damage. The choice and modification of the apparatus is thus at least as important as the chemical reaction per se. Thus, green chemistry can't make without in depth consideration of the engineering aspects. As a matter of fact, twelve chemical engineering principles have been formulated in parallel to those for green chemistry [2].[1]

1. *Inherent Rather Than Circumstantial*
   Designers need to strive to ensure that all materials and energy inputs and outputs are as inherently nonhazardous as possible, with attention to any potential issue, both physical properties such as viscosity and compressibility that affect heat and mass transfer, as well as explosivity, toxicity to humans and other organisms. This principle demands for the best possible design, that is that mass and energy transfer are defined around conveniently chosen boundaries, from a single reaction to a whole industrial park, and that the least environmentally unfriendly materials are chosen [3].

2. *Prevention Instead of Treatment*
   It is better to prevent waste than to treat or clean up waste after it is formed. Maximizing the yield of the saleable products and minimizing the yield of byproducts is the goal. This is identical to the second green chemistry principle [4].

3. *Design for Separation*
   Separation and purification operations should be designed to minimize energy consumption and materials use. A typical example is the quite energy intensive distillation, which contributes heavily to the CO_2 footprint of bulk chemicals. Making distillation more efficient, integrating the reaction into the distillation process, or looking for more effective separation techniques (e.g. reverse osmosis for saline water) must be explored [5].

4. *Maximize Efficiency*
   Products, processes, and systems should be designed to maximize mass, energy, space, and time efficiency. The motto "beyond the bench" indicates that already at the small scale stage the requirements for carrying out at the next higher level must be considered and a further improvement can be obtained when the optimization issues are integrated, rather than focused to separated areas (mass, energy, space, time) [6].

---

[1]A shorter list has been also proposed, see Abraham M, Nguyen N (2004) Green engineering: defining principles-results from the Sandestin conference. Environmental Progress 22:233–236.

1. Engineer processes and products holistically, use systems analysis, and integrate environmental impact assessment tools.
2. Conserve and improve natural ecosystems while protecting human health and wellbeing.
3. Use life-cycle thinking in all engineering activities.
4. Ensure that all material and energy inputs and outputs are as inherently safe and benign as possible.
5. Minimize depletion of natural resources.
6. Strive to prevent waste.
7. Develop and apply engineering solutions, while being cognizant of local geography, aspirations, and cultures.
8. Create engineering solutions beyond current or dominant technologies; improve, innovate, and invent (technologies) to achieve sustainability.
9. Actively engage communities and stakeholders in development of engineering solutions.

5. *Output-Pulled Versus Input-Pushed*
   Products, processes, and systems should be "output pulled", rather than "input pushed". Economies of scale should be extended to small scale plants, by producing exactly what is needed and optimal energy distribution. "The need should pull the act of production, rather than the ease and cost of production driving the need" [7].

6. *Conserve Complexity*
   Embedded entropy and complexity must be viewed as an investment when making design choices on recycle, reuse, or beneficial disposition. The chemical complexity of a product is a gift that must be conserved, not lost in subsequent steps. On the other hand, a high-level complexity makes products "less attractive for recycling and limits their second life to reuse only" [6].

7. *Durability Rather Than Immortality*
   Targeted durability, not immortality, should be a design goal.

8. *Meet Need, Minimize Excess*
   Design for unnecessary capacity or capability (e.g., "one size fits all") solutions should be considered a design flaw. This concept is strictly related to that of sustainable development.

9. *Minimize Material Diversity*
   Material diversity in multicomponent products should be minimized to promote disassembly and value retention.

10. *Integrate Material and Energy Flows*
    Design of products, processes, and systems must include integration and interconnectivity with available energy and materials flows. Thus, in heat exchanger networks (HEN) warm water from refrigerants is used to heat cold streams and in Mass Exchange Networks (MEN) lean streams are exploited for recovering mass from concentrated streams [8].

11. *Design for Commercial "Afterlife"*
    Products, processes, and systems should be designed for performance in a commercial "afterlife", in order to avoid piling up of waste.

12. *Renewable Rather Than Depleting*
    Material and energy inputs should be renewable rather than depleting the limited amount of resources available on Earth [9].

6.2 Progress Intensification

The application of several methods to the improvement of a process may result in a strategy called "*Process Intensification*" (PI), which had originally been implemented for the best use of the space available in a plant, but has now a larger definition. This refers to devising cheap and sustainable technologies that (potentially) result in a *dramatic* improvement of the present state of art by increasing the productivity, and/or lowering energy consumption and waste production [10]. This goal is obtained by introducing either new equipments or new methods as well as,

Table 6.1 Pros and cons of batch and continuous processing (reprinted with permission from Ref. [11])

	Batch processing	Continuous processing
PROS	Easy verification of product quality Regulations Suited to the shifts change Easier in case of low volume production Operations like crystallization, drying, precipitation are easily done in batch Flexible and versatile—important for small volume productions	Reduced plant size; high processing efficiency; reduced manufacturing costs; easier scale up; better process control—temperature, mass transfer, quality Unusual operating conditions (novel process windows)—reactions with highly concentrated or explosive mixtures made possible Reduced inventory, waste and energy requirements
CONS	Difficult scaling up Difficult to achieve homogeneous process conditions—changeable temperature, velocity and concentration profiles inside the equipment Poor mixing Safety, health and environmental issues	Complex to change processing mode once license has been obtained Difficult processing of solids-clogging, difficult cleaning Time and money intensive adaptation to strict batch-favored regulations Not cost effective in case of small scale applications when there already is in-ground capital for batch

most importantly, by the simultaneous development of both issues. As pointed out by Stankiewicz and Moulijn [10a], the key areas of development are the integration of more unit operations "into *multifunctional* reactors, the development of new *hybrid separations*, and the use of alternative forms and sources of *energy*" for carrying out the reactions.

Integration is more easily attained in reactions in flow, which are in fact increasingly adopted, and use mesoreactors (internal diameter of a few mm) or microreactors (o. d. <1 mm) in apparatuses measuring some centimeter to a meter overall. A comparison between flow and batch processes is drawn in Table 6.1 [11]. A minimization of residence time and waste production and a systematic control of the reaction course are obtained in this way (see Fig. 6.1). The small dimensions facilitate mass and heat transfer and under these conditions transport hindrances are eliminated, so that the reaction runs under intrinsic kinetic conditions. In a conventional apparatus heat transfer may be the limiting parameter, and force to carry out the reaction at a lower rate, since hot spots would otherwise be generated and decrease selectivity as well as threaten safety [12]. On the contrary, operating under harsh conditions is possible in flow systems, where even exploiting *ex regime* or runaway reactions can be considered. Indeed, reducing the volume of hazardous chemicals and continuously removing heat during the process make the processes intrinsically safer, even when strongly exothermic reactions (e.g. nitration of aromatics on one hand [13], photochemical reactions on the other) are carried out. Furthermore the modularity of flow reactors helps in adapting them to the actual process considered.

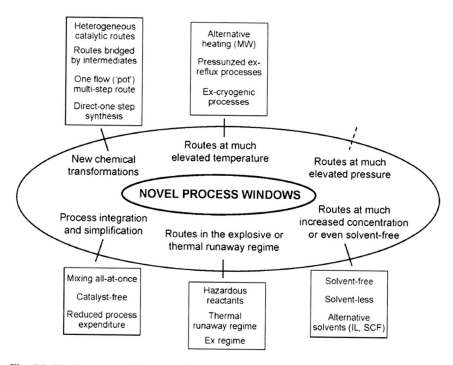

Fig. 6.1 Novel process window approach. Reprinted with permission from Ref. [15]

The improvement becomes much larger when both the chemical process and the way it is carried out are varied, not giving as established what found in previous investigations and having a predetermined maximum efficiency as the only determining facet. This opens a *New Window* of development, based on intensifying rather than on optimizing [14, 15] (some PI techniques are summarized in Fig. 6.1).

Most importantly, different operations can be combined in a single reactor. Examples are *reactive separation*, viz. the combination of chemical reaction and purification in a single column [12b] and *reactive distillation* [12c] probably the most common (and time honored) technique belonging to this category. In this case, the reactor functions also a still and an increased yield of the process due to the overcoming of thermodynamic equilibrium limitations by continuously subtracting one of the products (e.g. an alcohol during a transesterification process), the improved selectivity due to the limitation of undesired consecutive reactions and the reduced energy consumption in the case of exothermic reactions, since heat integration can be implemented [16]. It has been asserted that unit operation apparatuses may disappear, because they will become too expensive and inefficient for adoption in commercial applications, while of course remaining valuable for studying the steps of a process. There is no room to consider in detail the contribution of intensification to green chemistry, but a couple of examples for fuels, platform chemicals and APIs are reported below.

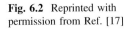

Fig. 6.2 Reprinted with permission from Ref. [17]

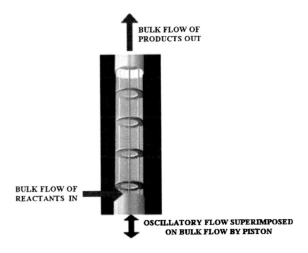

The production of biodiesel was strongly improved by the adoption of PI approach. Fatty acid methyl esters (FAME) based diesel were obtained through a NaOH catalyzed esterification of rapeseed oil and methanol in an oscillating flow reactor developed by Harvey et al. [17]. The device consisted of tubes containing equally spaced orifice plate baffles. An oscillatory motion was thus superimposed upon the net flow of the process fluid and the generated flow patterns caused an efficient heat and mass transfer. Under these conditions, a longer residence time was allowed and the apparatus was satisfactory for processes requiring such characteristic, as ester hydrolysis, usually limited to batch conditions [17] (Fig. 6.2).

Analogously, a thin film vortex fluid device was employed for the flow conversion of sunflower oil to FAME based biodiesel at room temperature. Optimization of the operating parameters avoided saponification of the starting oil, and high purity biodiesel was separated from glycerol byproduct and catalyst with no need for the otherwise conventional use of a co-solvent [18].

The use of supercritical solvents in the transesterification reaction has recently outmatched alkaline catalysis, since recover (or removal) of the catalyst from the crude biodiesel is avoided. As an example, Bertoldi et al. synthesized FAEE based diesel in supercritical ethanol/CO_2 mixture. Under these conditions solvent and produced esters form a single phase in the reactor and esterification occurs rapidly, avoiding product decomposition [19].

As for synthetic applications, Hessel [15] has summarized typical cases where Novel Process Windows conditions have been identified and a dramatic improvement has been obtained. Thus, fluorination of aromatics is usually carried out indirectly, e.g. via the Balz-Schieman reaction involving nitration, reduction, diazotization and decomposition of the diazonium salts. *Direct* fluorination by using elemental fluorine is however accessible in a falling film reactor. A <1 μm liquid layer was formed and the large gas-liquid surface ensured efficiency and safety. Microfabricated reactors were used for determining the dependence of the reaction on the flux regime. A good selectivity at significant conversion was obtained (e.g. in the case of toluene in

Scheme 6.1 Multicomponent synthesis of diketone **1**

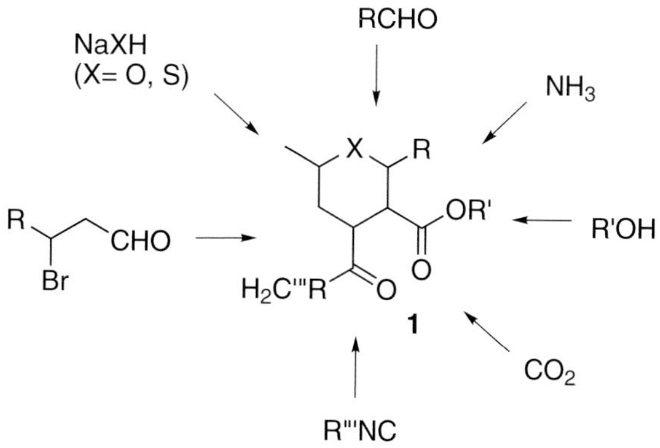

acetonitrile a combined selectivity of the three ring-substituted monofluorotoluenes of up to 24 % at 77 % conversion was obtained) [20, 21].

Flow processes have been successfully applied to multicomponent reactions, exploiting the intrinsic complementary reactivity of such processes. Thus, complex molecule **1** has been prepared via the Ugi route by introducing the complete mixture of reactants at the flow entry. A library of target compounds was obtained in a combinatorial fashion, by simply changing the structure of the introduced reactant (see Scheme 6.1) [22, 23].

Micro flow reactors are the best choice for performing processes involving reagents of limited stability, such as Grignard salts or alkyl lithium derivatives. As an example, alcohol **2** was obtained in kg scale by treating 3-methoxyphenyllithium with cyclohexanone at low temperature (see Scheme 6.2) [24].

Another case of a strong enhancement of selectivity in a microflow system has been reported for the derivatization of sugars. A silicon microreactor with a glass surface (Fig. 6.3) was fabricated by adopting the Deep Reactive Ion Etching technique, and used in glycosylation reactions by Ratner et al. [25]. The new device allowed to operate at a higher temperature (-35 °C) with a shorter residence time (25.7 s) as compared to 213 s at -60 °C for the batch procedure (see compound **3** in Scheme 6.3).

An efficient system for tris(trimethylsilyl)silane (TTMSS) mediated hydrosilylation reactions in superheated toluene at 130 °C was carried out in a microstructured device equipped with a backpressure valve. Conventional toxic or chlorinated

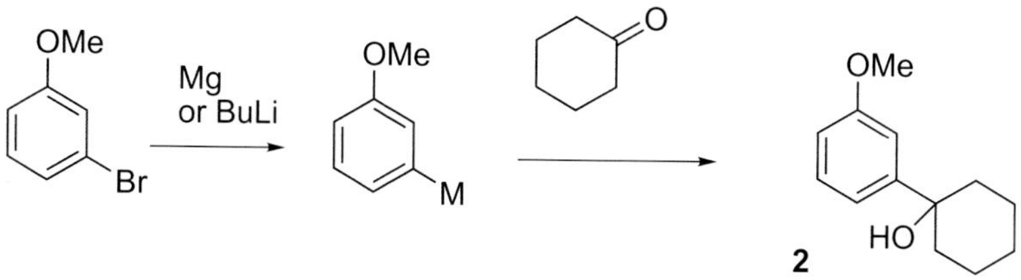

Scheme 6.2 Treating 3-methoxyphenyllithium with cyclohexanone at low temperature

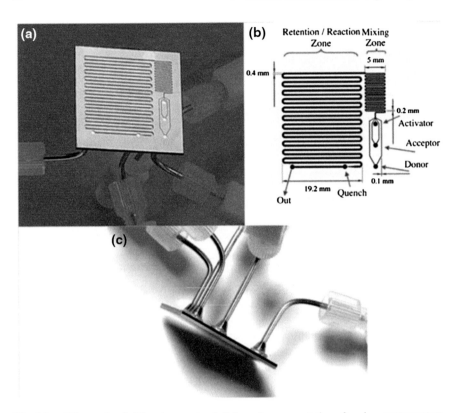

Fig. 6.3 a Silicon microfluidic microreactor. **b** Schematic representation of a microreactor system, comprised of three primary inlets, a mixing and reaction zone, a secondary inlet for quench, and an outlet for analysis/collection. **c** Soldered joints of microreactor, also perspective of device from side. Reprinted with permission from Ref. [25]

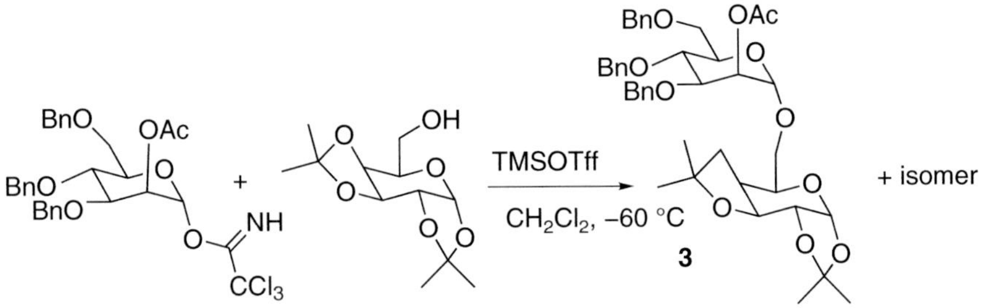

Scheme 6.3 Batch procedure for the derivatization of sugars

solvents (such as CCl$_4$ or benzene) were avoided in this procedure and the reaction proceeded with a an E/Z ratio larger than that observed in batch systems (see Scheme 6.4, compound **4**) [26].

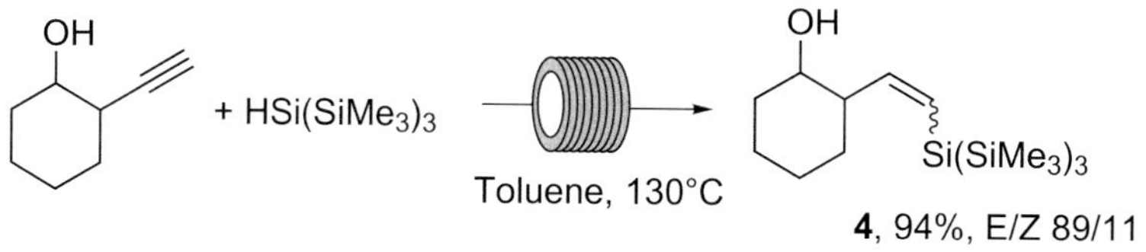

Scheme 6.4 Tris(trimethylsilyl)silane (TTMSS) mediated hydrosilylation reactions

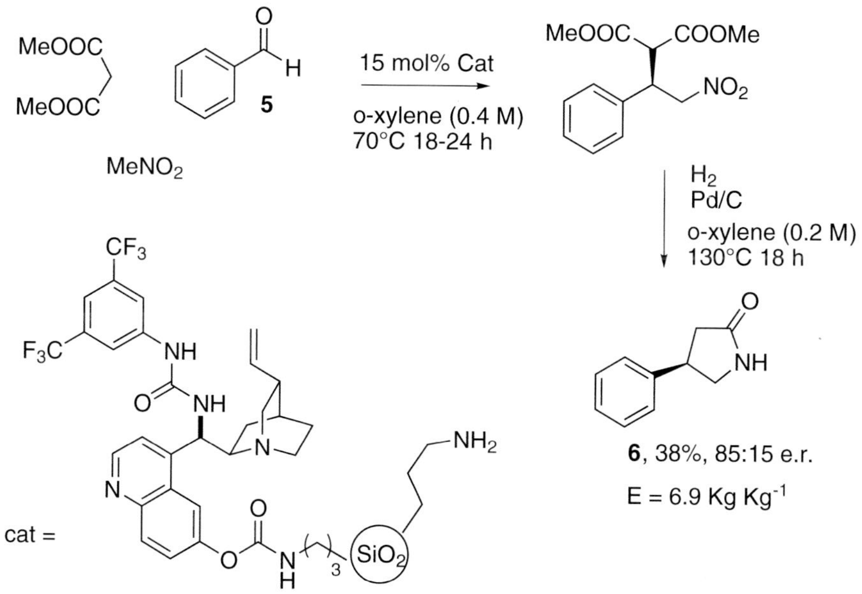

Scheme 6.5 One pot synthesis of γ-aminobutyric acid GABA derivative **2**

Multifunctional materials able to perform one-pot multistep reactions are one of the open fronts in catalysis. Multisite organic–inorganic hybrid catalysts have been prepared and applied in a new general, practical, and sustainable synthetic procedure toward industrially relevant γ-Aminobutyric acid (GABA) derivatives **6**. The domino sequence is composed of seven chemical transformations which are performed in two one-pot reactions [27]. The synthetic route involved a multicomponent reaction between aldehyde **5**, nitromethane and dimethylmalonate, catalyzed by commercially available solid catalysts. The use of an urea-modified cinchona alkaloid derivatives linked on a microporous siliceous material with additional pending aminopropyl groups allowed the synthesis of the products in both enantiomeric forms after a single column purification step.

The raw product was then transferred to the second reactor for the heterogeneous catalytic hydrogenation of the nitro group to the primary amine, which underwent cyclization to amide. Ensuing thermal decarboxylation of the remaining ester group gave the desired final product (Scheme 6.5) [27].

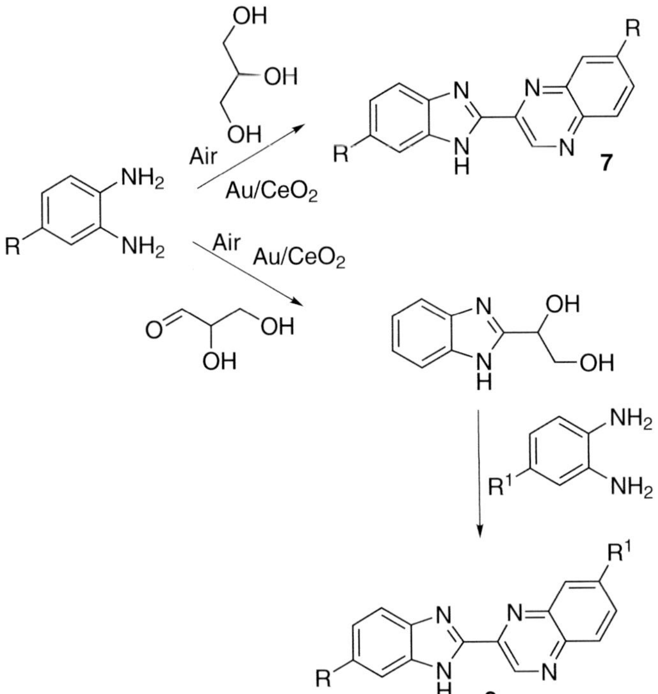

Scheme 6.6 Synthesis of substituted benzimidazoylquinoxalines [28]

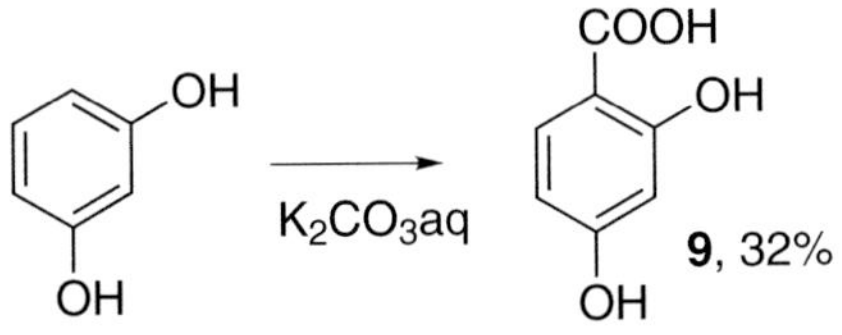

Scheme 6.7 Synthesis of β-resorcylic acid **9**

Gold catalysis has been used for the synthesis of benzimidazolylquinoxalines **7, 8** from glycerol derivatives (Scheme 6.6) in a multistep one-pot methodology. Thus, an oxidation–cyclization of glycerol with o-phenylenediamine derivatives afforded benzimidazoylquinoxaline compounds with the same substituents in both heterocycles, while products with different substituents in each aromatic ring were obtained via coupling of o-phenylenediamine derivatives with glyceraldehyde and oxidative cyclization of the obtained benzimidazole with another o-phenylenediamine in a second step. In each stage gold nanoparticles supported on nanoparticulated ceria (Au/CeO$_2$) was the catalyst and air the oxidant (Scheme 6.6) [28].

The continuously operated Kolbe-Schmitt synthesis of 2,4-dihydroxybenzoic acid (β-resorcylic acid, **9**. Scheme 6.7) was subjected to process intensification in the field

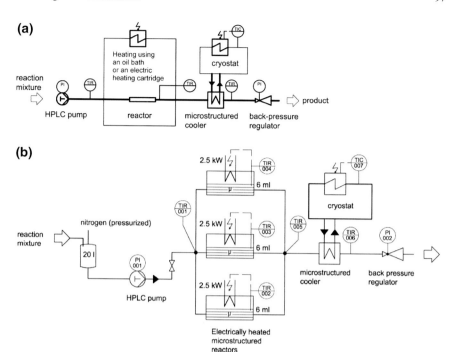

Fig. 6.4 Apparatus for the carboxylation of resorcinol to **5**. **a** A laboratory setup for the use of an oil bath-heated 1/16-in. capillary reactor or an electrically heated microstructured reactor. **b** Flow diagram of the pilot plant using three electrically heated microstructured reactors operated in parallel. Reprinted with permission from Ref. [29]

of Novel Process Windows cluster of the Deutsche Bundesstiftung Umwelt (Fig. 6.4). The approach developed was based on a phased approach that involved: (a) process transfer of the examined reaction from batch to continuous operation; (b) flow-process optimization; (c) examination of alternative process options; (d) scale-up of the optimized process up to pilot scale; (e) transferring the results of the model reaction to other reactions. In a first step, the treatment of resorcinol with aqueous potassium hydrogen carbonate was moved from the capillary reactor to a tailor-made electrically heated microstructured reactor equipped with 40 microchannels [29]. This enabled a 25-fold increase of the capacity. The same reaction was then successfully tested in a pilot plant comprising three of such reactors operated in parallel. The benefits of this approach are well highlighted by the lowering of the reaction times from more than 2 h down to 6 s, achieving a space time yield of 28,800 kg m^{-3} h^{-1} for the pilot scale. Moving from the capillary (9 g h^{-1}) to the pilot reactor, the productivity could be increased up to 520 g h^{-1} [29].

Process intensification is in fact increasingly used, particularly for APIs [30–32], where green metrics play a key role in analyzing and optimizing the process. In particular, LCA has been applied as a multiple criteria decision analysis within

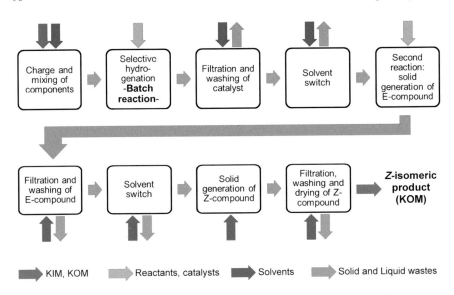

Fig. 6.5 Approach adopted for obtaining the intensification of an API. Reprinted with permission from Ref. [31]. *KIM* Key imput material, *KOM* Key output material

pharmaceutical process optimization and intensification. An API produced at Sanofi was chosen as an appropriate example for the relatively low amount produced (scale, 100 kg year^{-1}) and the price (a few thousand euros). This made worthwhile to use the more sophisticated LCA approach rather than mass green metrics such as PMI generally employed for APIs, in order to obtain a holistic examination [33]. The production is based on the chemo and stereo selective reduction of aromatic nitro groups, in the presence of alkene functionalities to yield the respective Z- and E-amino compounds, the former being the economically valuable one. Transfer from batch to continuous processing was considered along with both already known and hypothetical alternative catalytic systems. The overall inventory analyzed is indicated in Fig. 6.4 [30].

The environmental impact of various already known as well as hypothetical process options was evaluated in order to identify bottlenecks and improvement potentials for further process development activities. The strategy depicted in Fig. 6.5 was followed and the chance to significantly decrease of the CO_2 imprint value (765 kg CO_2 equivalents) was pointed out. A similarly analysis was carried out for process intensification of other APIs [31] (see Fig. 6.6).

A feeling of the problem confronted can be gathered from the analysis of the synthesis of Galantamine·bromhydrate **10**. The first generation production pathway for this API consisted of nine consecutive steps. Analysis of this process evidenced that 46 pieces of equipment were required and overall 485 basic operations were

Fig. 6.6 Approach adopted for obtaining the intensification of an API. Reprinted with permission from Ref. [31]

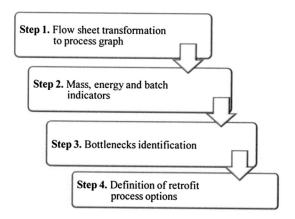

performed. For each of these 485 basic operations a mass and energy balance was set up at three levels:

- at the process level (or equipment level including reactor, filter, dryer etc.).
- at the gate-to-gate level (taking into account waste treatment such as solvent distillation, waste water treatment, VOC recovery, etc., as well as storage and utility production such as heating, cooling, purified water, pressurized air etc.).
- at the cradle-to-gate level by combining all data of the gate-to-gate level with life cycle inventory data.

The mass and energy balances of all such operation were converted in terms of energy and expressed in Joules. In this way, the "irreversibilities at the process level, the energy losses at the gate-to-gate level and the cumulative energy extracted from the natural environment (CEENE) could be quantified for the production of each intermediate" (A till H in Fig. 6.5) and the final API. For the last one the energy loss was expressed as CEENE per mol API. The synthetic plans were compared. In the second generation process, the fourth and fifth of the nine steps of the first generation were optimized by replacing a solvent and by improving the efficiency of both steps.

In the third generation of the process pathway a continuous process in a flow reactor was introduced replacing the original sixth steps by a continuous process using a flow reactor. This comparison showed that the overall resource consumption from the first to the third generation decreased by 41 % due to increased resource efficiency (see Fig. 6.7) [33].

Ionic liquids (ILs) have been exploited in various areas of research at laboratory scales, suggesting the adoption, in the future of ILs-based industrial process. In any case, the limiting issue is the high cost of most laboratory-synthesized ILs which limit application to small scale processes. However, ILs can be also inexpensive. The economic advantages obtained by intensification processes for the synthesis of APIs in hydrogen sulfate based ionic liquids through an acid–base neutralization

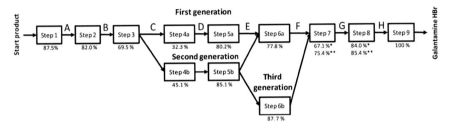

Fig. 6.7 Improving the synthesis of Galantamine·* HBr [33]

were pointed out by Hallet et al. [34]. The cost of triethylammonium hydrogen sulfate was quantified in 1.24 \$ kg^{-1}, which compared favorably with typical organic solvents such as acetone or ethyl acetate. In particular, although no significant improvement of the yield has been observed, a general lowering of the costs related to the use of cooling water and electricity was quantified.

As hinted in Chap. 4, platform chemicals levulinic acid and its esters **11** are currently prepared by treatment of cellulose and hemicellulose biomass with inorganic acids. Alternatively, these materials and derivatives are prepared by acid-catalyzed hydrolysis or alcoholysis of furfuryl alcohol, in turn obtained from glucose and xylose containing biomass and thus through three separate steps. A mesoporous ZrNb binary phosphate solid acid (cat in scheme 6.8) with high surface area (170.1 m^2 g^{-1}) was synthesized by a sol–gel method and used as support for Pt nanoparticles. This catalyst was employed in an integrated microsystem for the one-pot conversion of furfural to levulinate ester (Scheme 6.8) [35].

The extensive work of Buchwald et al. in the field of coupling reactions involving in situ generated aryl triflates was focused on avoiding time consuming isolation of intermediate substrates. A microreactor where both Heck and Suzuki reactions were carried out was developed and triflates were obtained in situ by treatment of the corresponding phenols **12** with triflic anhydride in the presence of an amine as a base. The reaction mixture was thus washed with aqueous hydrochloric acid, extracted in a microseparator. The crude triflate undergoes Suzuki coupling in a water–toluene mixture and the resulting biaryl **13** was obtained in excellent yields (ranging from 83 % up to quantitative) with an average resident reaction time of 400 s for both steps [36] (see Scheme 6.9).

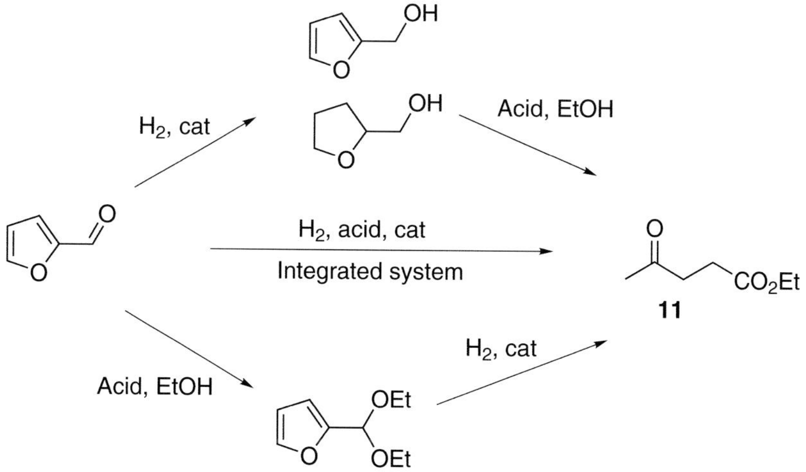

Scheme 6.8 One pot synthesis of ethyl levulinate **11**

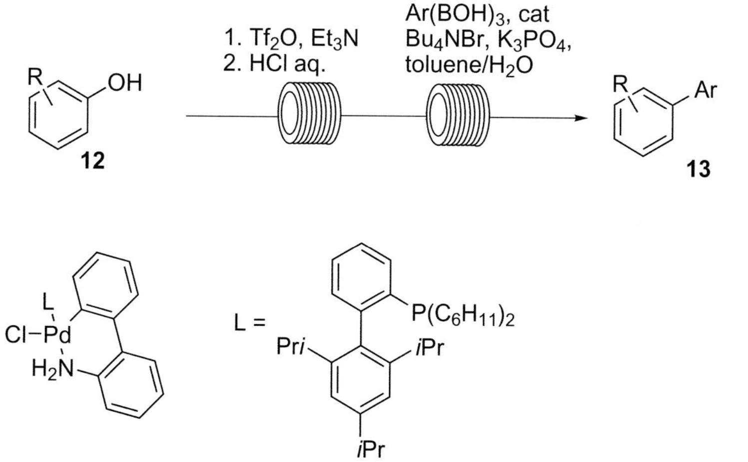

Scheme 6.9 In situ arylation of phenols [36]

References

1. Adriano DC (2001) Trace elements in terrestrial environments. Biogeochemistry, bioavailability, and risks of metals. In: Daughton CG, Jones-Lepp TL (eds) Pharmaceuticals and personal care products in the environment: scientific and regulatory issues, ACS symposium series 791, American Chemical Society, Washington. Springer, New York
2. Anastas, PT, Zimmerman JB (2003) Design through the twelve principles of green engineering. Env Sci Tech 37:94A–101A. See also Garcìa-Serna J, Pérez-Barrigòn L,

Cocero MJ (2003) New trends for design towards sustainability in chemical engineering: green engineering. Chem Eng J 133:7–30
3. Constable DC. In http://www.acs.org/content/acs/en/greenchemistry/what-is-green-chemistry/principles.html
4. Abraham M in Ref. [3]
5. Reallf MJ, Wang D in Ref. [3]
6. Gonzalez MA in Ref. [3]
7. Mattews MA in Ref. [3]
8. Jiménez Gonzàlez C in Ref. [3]
9. Anastas PT, Zimmerman JB (2003) Design through the twelve principles of green engineering. Env Sci Tech 37:94A–101A
10. (a) Stankiewicz AI, Moulijn JA (eds) (2004) Reengineering the chemical processing plant. Process intensification. Marcel Dekker, New York. (b) Ramshaw C (1985) Process intensification: heat and mass transfer. Chem Eng 415:30–33
11. Denčić I, Ott D, Kralisch D, Noël T, Meuldijk J, de Croon M, Hessel V, Laribi Y, Perrichon P (2014) Eco-efficiency analysis for intensified production of an active pharmaceutical ingredient: a case study. Org Proc Res Dev 18:1326–1338
12. (a) Kumar V, Nigam KDP (2012) Process intensification in green synthesis. Green Process Synth 1:79–107. (b) For different process intensification techniques, see for reviews: Stankiewicz A (2003) Reactive separations for process intensification: an industrial perspective. Chem Eng Process 42:137–144. (c) Malone MF, Huss RS, Doherty MF (2003) Green chemical engineering aspects of reactive distillation. Environ Sci Technol 37:5325–5329
13. See for instance Antes J, Boskovic D, Krause H, Loebbecke S, Lutz N, Tuercke T, Schweikert W (2003) Analysis and improvement of strong exothermic nitrations in microreactors. Chem Eng Res Des 81:760–765
14. BHR Group. www.bhrgroup.com/pi/aboutpi.htm. Keil FJ (ed) (2007) Modeling of process intensification. Wiley, Weinheim
15. Hessel V (2009) Novel process windows—gate to maximizing process intensification via flow chemistry. Chem Eng Technol 32:1655–1681
16. Noeres C, Kenig EY, Gòrak A (2003) Modelling of reactive separation processes: reactive absorption and reactive distillation Chem Eng Process 42:157–178
17. Harvey P, Mackley AP, Selinger MR (2003) Process intensification of biodiesel production using a continuous oscillatory flow reactor. J Chem Technol Biotechnol 78:338–341
18. Britton J, Raston CL (2014) Continuous flow vortex fluidic production of biodiesel. RSC Adv 4:49850–49854
19. Bertoldi C, da Silva C, Bernardon JP, Corazza ML, Filho LC, Oliveira JV, Corazza FC (2009) Continuous production of biodiesel from soybean oil in supercritical ethanol and carbon dioxide as cosolvent. Energy Fuels 23:5165–5172
20. Jähnisch K, Baerns M, Hessel V, Ehrfeld W, Haverkamp V, Löwe H, Wille C, Guber A (2000) Direct fluorination of toluene using elemental fluorine in gas/liquid microreactors. J Fluorine Chem 105:117–128
21. de Mas N, Günther A, Schmidt MA, Jensen KF (2003) Microfabricated multiphase reactors for the selective direct fluorination of aromatics. Ind Eng Chem Res 42:698–710
22. Ugi I, Almstetter M, Gruber B, Heilingbrunner M (1997) MCR XII. Efficient development of new drugs by online-optimization of molecular libraries. Springer, Berlin, pp 190–194
23. Mitchell MC, Spikmans V, Bessot F, Manz A, de Mello A (2000) Towards organic synthesis in microfluidic devices: multicomponent reactions for the construction of compound libraries. Kluwer, Dordrect, pp 463–465
24. Zhang X, Stefanick S, Villani FJ (2004) Application of microreactor technology in process development. Org Proc Res Dev 8:455–460
25. Ratner DM, Murphy ER, Jhunjhunwala MG, Snyder DA, Jensen KF, Seeberger PH (2005) Microreactor-based reaction optimization in organic chemistry-glycosylation as a challenge. Chem Commun 578–580

26. Odedra A, Geyer K, Gustafsson T, Gilmour T, Seeberger PH (2008) Safe, facile radical-based reduction and hydrosilylation reactions in a microreactor using tris(trimethylsilyl)silane. Chem Commun 3025–3027

27. Leyva-Pérez A, Garcia PG, Corma A (2014) Multisite organic–inorganic hybrid catalysts for the direct sustainable synthesis of GABAergic drugs. Angew Chem Int Ed 53:8687–8690

28. Climent MJ, Corma A, Iborra S, Martinez-Silvestre S (2013) Gold catalysis opens up a new route for the synthesis of benzimidazoylquinoxaline derivatives from biomass-derived products (glycerol). ChemCatChem 5:3866–3874

29. Krtschil U, Hessel V, Kost HJ, Reinhard D (2013) Kolbe-Schmitt flow synthesis in aqueous solution—from lab capillary reactor to pilot plant. Chem Eng Technol 36:1010–1016

30. Ott D, Kralisch D, Dencic I, Hessel V, Laribi Y, Perrichon PD, Berguerand C, Kiwi-Minsker L, Loeb P (2014) Life cycle analysis within pharmaceutical process optimization and intenfication: case study of active pharmaceutical ingredient. ChemSusChem 7:3521–3533

31. Kralisch D, Ott D, Gericke D (2015) Rules and benefits of life cycle assessment in green chemical process and synthesis design: a tutorial review. Green Chem 17:123–145

32. Boodhoo K, Harvey A (eds) (2013) Process intensification for green chemistry. Engineering solutions for sustainable chemical processing. Wiley, New York. Renken A (2014) Process intensification for clean catalytic technology. In: Wilson K, Lee AF (eds) Heterogeneous catalysts for clean technology, pp 333–364

33. Van der Vorst G, Aelterman W, De Witte B, Heirman B, Van Langenhove H, Dewulf J (2013) Reduced resource consumption through three generations of Galantamine·HBr synthesis. Green Chem 15:744–748

34. Long Chen L, Sharifzadeh M, Mac Dowell N, Welton T, Shahc N, Hallett JP (2014) Inexpensive ionic liquids: $[HSO_4]^-$-based solvent production at bulk scale. Green Chem 16:3098–3106

35. Chen B, Li F, Huang Z, Lu T, Yuan Y, Yuan G (2014) Integrated catalytic process to directly convert furfural to levulinate ester with high selectivity. ChemSusChem 7:202–209; see also: Sanders JPM, Clark JH, Harmsenc GJ, Heeres HJ, Heijnend JJ, Kerstene SRA, van Swaaije WPM, Moulijn JA (2012) Process intensification in the future production of base chemicals from biomass. Chem Eng Process 51:117–136

36. Noël T, Kuhn S, Musacchio AJ, Jensen KF, Buchwald SL (2011) Suzuki–Miyaura cross-coupling reactions in flow: multistep synthesis enabled by a microfluidic extraction. Angew Chem 123:6065–6068. For recent reviews on flow synthesis see: Mason BP, Price KE, Steinbacher JL, Bogdan AR, McQuade DT (2007) Greener approaches to organic synthesis using microreactor technology. Chem Rev 107:2300–2318. Vaccaro L, Lanari D, Marrocchi A, Strappaveccia G (2014) Flow approaches towards sustainability. Green Chem 16:3680–3704. Malet-Sanz L, Flavien Susanne L (2012) Continuous flow synthesis. A pharma perspective. J Med Chem 55:4062–4098. Gutmann B, Cantillo D, Kappe CO (2015) Continuous-flow technology—A tool for the safe manufacturing of active pharmaceutical ingredients. Angew Chem Int Ed 54:6688–6728

Chapter 7
Conclusions and Outlook

Adopting the green chemistry perspective in every process carried out is not simple, actually this requires choosing among the many possibilities the "greenest" method and thus taking advantage of any advancement in the knowledge of chemical reactions. Furthermore, the interplay between methods typical of the industry practice and those typical of university labs is particularly important in this case. Industry carries out an in-depth development of a limited number of processes. Sometimes, this involves methods requiring a sophisticated apparatus that are not easily carried out in a university lab, and the in depth optimization of a catalyst may be less easily performed in a research lab. However, the advanced development of catalysts for a particular industrial use may result in materials highly active also for applications different from the one initially considered. An example is found in zeolite-based catalysts that have been studied quite intensively due to their role in oil refining, but have then found a plethora of valuable applications in both small and large scale procedures for different processes. A main direction in the development of this discipline is given by *biocatalysis* that has been more and more applied to fine and bulk chemicals production (see Chaps. 3 and 4).

On the other hand, less common activation methods, such as *microwave* irradiation or *photochemistry* (Chap. 3), have been easily introduced in a research lab, but require an extensive rethinking of the apparatus, and thus may encounter severe limitations at the scaling up stage, which makes such methods less palatable for the industry. Green chemistry demands that all of the possibilities are considered and their environmental performance is assessed. For this reason it is important that the paradigms of green chemistry (*the use of chemicals deriving from renewable sources, the adoption of mild activation methods for the synthesis, the use of environmental friendly solvents, the introduction of green metrics in order to enable to detect and quantify any negative effect*) become familiar among students in order that a new attitude is adopted already at the education level. Admittedly, many aspects are still far from a satisfactory formulation. As an example, green metrics has seen a rapid development with the blossoming of a variety of proposals (see Chap. 2) that, however, make the field seem somewhat confused. At present, the most used method is the simplest one, the E-factor that yields a limited information, but is easily calculated. On the other extreme, sophisticated LCA analyses

A. Albini and S. Protti, *Paradigms in Green Chemistry and Technology*,
SpringerBriefs in Green Chemistry for Sustainability,
DOI 10.1007/978-3-319-25895-9_7

are being carried out in an increasing number and have demonstrated to be fully worthwhile when an actual optimization is considered.

A great step forward would be that handbooks of organic chemistry report at least the E-factor or the PMI value for every reaction quoted, making students familiar from the beginning with the consideration of all the aspects of a reaction, not only the "desired" process as usually indicated. In the same way, indication of the environmental performance and of the energetic requirement should be given for every new reaction, so that this key aspect of synthesis is easily available to chemists.

Some advancement in this direction has been effected, as in the case of the Neue Organische Praktikum [1], where such an approach has been adopted for the (limited) number of reactions considered and students profit of this perspective and better appreciate the correct attitude in the lab. Likewise, green characteristics in research work can be evidenced, e.g. by using an electronic notebook such as the Perkin Elmer E-Notebook, where dedicated features have been introduced for this purpose [2a], so that new finding are made available to the chemical community as soon as possible. Green Chemistry is considered as a challenge in synthesis also by the Organic Chemistry Portal, in which an almost exaustive collection of procedures is included [2b]. The preparation of "Guides" for the choice of reagents and solvents is another welcome addition that will surely acquire more importance while more data ad more critical contributions are added [3].

With a shade of optimism, it appears reasonable to think that green chemistry will become an inseparable part of everyday chemistry and thinking the green way will become a natural habit for every chemistry practitioner. Science is what scientists think and an early contact with green chemistry will early result in a clean, environment respectful (and profitable) chemistry [4–6].

The transformation of a set of inalienable moral principles into a set of practical paradigms will then be completed and a sustainable future will become an actual possibility.

References

1. http://www.oc-praktikum.de/nop/en-article-why
2. (a) http://blog.cambridgesoft.com/post/2013/01/09/Green-chemistry-made-easy-simple-configurations-for-ELN.aspx. (b) see http://www.organic-chemistry.org/
3. Henderson RK, Jiménez-Gonzàlez C, Constable DJC, Alston SR, Inglis GGA, Fisher G, Sherwood J, Binks, SP, Curzons AD (2011) Expanding GSK'solvent selection guide—embedding sustainability into solvent selection starting at medicinal chemistry. Green Chem 13:854–862. (b) Adams JP, Alder CM, Andrews I, Bullion AM, Campbell-Crawford M, Darcy MG, Hayler JD, Henderson RK, Oare CA, Pendrak I, Redman AM, Shuster LE, Sneddon HF, Walker MD (2013) Development of GSK's reagent guides—embedding sustainability into reagent selection. Green Chem 15:1542–1549. (See Fig. 1 in Chap. 3 and Fig. 2 in Chap. 5)
4. http://www2.epa.gov/green-chemistry/benefits-green-chemistry
5. Watson WJW (2012) How do the fine chemical, pharmaceutical, and related industries approach green chemistry and sustainability? Green Chem 14:251–259
6. Ricci A, Albini A (2014) Green and sustainable. Speciality Chemicals 34:14–17

Index

© The Author(s) 2016
A. Albini and S. Protti, *Paradigms in Green Chemistry and Technology*,
SpringerBriefs in Green Chemistry for Sustainability,
DOI 10.1007/978-3-319-25895-9

Printed in the United States
By Bookmasters